THE U.S. GOVERNMENT
& RENEWABLE ENERGY

Pan Stanford Series on Renewable Energy

Series Editor
Wolfgang Palz

Vol. 1
Power for the World: The Emergence of Electricity from the Sun
Wolfgang Palz, ed.
2010
978-981-4303-37-8 (Hardcover)
978-981-4303-38-5 (eBook)

Vol. 2
Wind Power for the World: The Rise of Modern Wind Energy
Preben Maegaard, Anna Krenz, and Wolfgang Palz, eds.
2013
978-981-4364-93-5 (Hardcover)
978-981-4364-94-2 (eBook)

Vol. 3
Wind Power for the World: International Reviews and Developments
Preben Maegaard, Anna Krenz, and Wolfgang Palz, eds.
2013
978-981-4411-89-9 (Hardcover)
978-981-4411-90-5 (eBook)

Vol. 4
Solar Power for the World: What You Wanted to Know about Photovoltaics
Wolfgang Palz, ed.
2013
978-981-4411-87-5 (Hardcover)
978-981-4411-88-2 (eBook)

Vol. 5
Sun above the Horizon: Meteoric Rise of the Solar Industry
Peter F. Varadi
2014
978-981-4463-80-5 (Hardcover)
978-981-4613-29-3 (Paperback)
978-981-4463-81-2 (eBook)

Vol. 6
Biomass Power for the World: Transformations to Effective Use
Wim van Swaaij, Sascha Kersten, and Wolfgang Palz, eds.
2015
978-981-4613-88-0 (Hardcover)
978-981-4669-24-5 (Paperback)
978-981-4613-89-7 (eBook)

Vol. 7
The U.S. Government & Renewable Energy: A Winding Road
Allan R. Hoffman
2016
978-981-4745-84-0 (Paperback)
978-981-4745-85-7 (eBook)

Forthcoming

Vol. 8
Sun towards High Noon: Meteoric Rise of the Solar Industry Continues
Peter F. Varadi
2017

Pan Stanford Series on Renewable Energy
Volume 7

THE U.S. GOVERNMENT & RENEWABLE ENERGY

A Winding Road

Allan R. Hoffman

PAN STANFORD PUBLISHING

Published by

Pan Stanford Publishing Pte. Ltd.
Penthouse Level, Suntec Tower 3
8 Temasek Boulevard
Singapore 038988

Email: editorial@panstanford.com
Web: www.panstanford.com

British Library Cataloguing-in-Publication Data
A catalogue record for this book is available from the British Library.

The U.S. Government & Renewable Energy: A Winding Road
Copyright © 2016 Allan R. Hoffman

ISBN 978-981-4745-84-0 (Paperback)
ISBN 978-981-4745-85-7 (eBook)

Printed in the USA

This book is dedicated to my wife, who has had to patiently listen to me talk about energy issues and the inevitable transition for more than 30 years and to my children and grandchildren and their peers who will inherit the energy systems we helped create.

I'd put my money on the sun and solar energy.
What a source of power! I hope we don't have to wait
till oil and coal run out before we tackle that.
I wish I had more years left!
—Thomas Edison, 1931

Contents

Preface

Why another book on energy? It is not a book on the details of energy technologies but rather on how the United States and other governments have changed their thinking about energy issues over the past four decades. This change has been triggered by increasing concern about the role of fossil fuels in global warming and climate change, greater awareness of the risks of nuclear power, and the emergence of viable renewable energy sources.

It did not come easily, especially in the United States, where traditional energy industries argued that renewable energy was too expensive and unable to meet U.S. energy needs, and politicians dependent on energy industry contributions supported the status quo through legislative inaction. It was a shortsighted approach that has hurt the U.S. economy by allowing other countries to take the lead and reap the related job and economic growth benefits.

One contribution of this book will be to help people understand how this change came about in the United States from the perspective of a well-placed participant and observer. To borrow a phrase I came across recently in a book review, the book will offer "an intimate, novelistic sense of what it was like to be there." Another purpose is to enhance understanding of the global energy transition that is finally getting underway. It has been long in coming and validates my long-held belief and that of many others that the world energy system must undergo an inevitable transition from heavy dependence on fossil fuels following the advent of the industrial revolution in the 1800s to an emerging energy system that will rely increasingly on renewable energy. The second decade of the 21st century has seen the beginnings of this transition, which is now unfolding at an accelerating, even dizzying, pace.

The book reflects my personal experiences and observations from when I first began to educate myself about energy efficiency and renewable energy starting in 1969, and then my move to Washington, DC, in 1974 to learn about how our federal government was handling these issues. What started out as a one-year leave of absence from my academic position as a young physics professor, via a fellowship to work with Congress, turned out to be a life-changing event that turned me into that most dreaded of terms, a bureaucrat.

The book also describes how I first got interested in energy issues, my early experiences in New England as the national debate on nuclear power got underway, my moving to Washington in 1974 to work as a Congressional Fellow and Staff Scientist for the U.S. Senate Commerce Committee, my move in 1978 to the newly formed Department of Energy (DOE) as a political appointee in the Carter Administration, my time in the 1980s at the National Academies of Sciences, and my subsequent return in 1991 to the executive branch to serve in various senior positions at the DOE. It also describes my efforts to describe the critical linkage between energy and water issues, which is today attracting considerable attention.

Nevertheless, the book's major purpose is to illustrate how the U.S. government moved along its winding path to where it is today in getting ready for a renewable energy future. It is important to know where we came from as we decide where we go. Target audiences are the young people who will inherit the transition and shape its future. Other target audiences are those in government who currently shape our public policies and those colleagues, friends, and family members who lived through many of the times and events discussed in the book. It has been an exciting ride that has hopefully left positive marks along the way and a positive legacy for future generations.

Acknowledgments

This book and the history it relates did not occur in a vacuum. Over the past 40 years, I have worked with many people to whom I am grateful for invaluable discussions, useful and accurate information, informed insights, loyalty, hard work and long hours, and sound advice. They are too numerous to list in this document, but I know who they are and I say thank you from the bottom of my heart. It's been a good ride, made possible by all your contributions.

I do want to single out one individual who has become a good friend in recent years and has guided me through the trials and tribulations of writing and publishing a book, something he has solid experience with. That is Dr. Peter F. Varadi, one of the true solar energy pioneers who helped put us on the renewable energy path. Thank you, Peter.

Glossary

Please use this Glossary as a guide to the many abbreviations (acronyms) you will find throughout the text of this book. It reflects my too many years in Washington, DC, where abbreviations are often not an insignificant fraction of the language of daily communication. I have spelled out acronyms the first time they appear in the text but not after that, so this Glossary may often come in handy.

- AAAS: American Association for the Advancement of Science
- AC: alternating current
- ACOG: Atlantic Committee for the Olympic Games
- ACS: American Chemical Society
- AID: Agency for International Development
- APS: American Physical Society
- ASME: American Society of Mechanical Engineers
- BIPV: building-integrated photovoltaics
- CEQ: Council on Environmental Quality
- COSEPUP: Committee on Science, Engineering, and Public Policy
- CSP: concentrating solar power
- CWEA: California Wind Energy Association
- DPR: Domestic Policy Review
- DC: District of Columbia (and direct current)
- DoD: Department of Defense
- DOE(1): Department of Energy
- DOE(2): Department of Education
- DOI: Department of Interior
- DOS: Department of State
- DOT: Department of Transportation
- DSIRE: Database of State Incentives for Renewables & Efficiency
- EERE: Energy Efficiency and Renewable Energy
- EPRI: Electric Power Research Institute
- ERDA: Energy Research and Development Administration

- ESTCP: Environmental Security Technology Certification Program
- EUROSOLAR: European Association for Renewable Energy
- FE: fossil energy
- FEA: Federal Energy Administration
- GM: General Motors
- MAOGA: Metropolitan Atlanta Olympic Game Authority
- MIT: Massachusetts Institute of Technology
- MRI: magnetic resonance imaging
- MWe: megawatts electrical
- MWp: megawatts peak
- NADA: National Automobile Dealers Association
- NAS: National Academy of Sciences
- NECNP: New England Coalition on Nuclear Pollution
- NERC: National Energy Research Center (Jordan)
- NHTSA: National Highway Traffic and Safety Administration
- NRC: National Research Council
- NREL: National Renewable Energy Laboratory
- NSF: National Science Foundation
- NSA: National Security Agency
- OMB: Office of Management and Budget
- ORNL: Oak Ridge National Laboratory
- OAS: Organization of American States
- OSW: offshore wind
- OTA: Office of Technology Assessment
- OTEC: ocean thermal energy conversion
- OUT: Office of Utility Technologies
- PG&E: Pacific Gas and Electric
- RANN: Research Applied to National Needs
- RFP: request for proposal
- ROWPU: reverse osmosis water purification unit
- RPI: Rensselaer Polytechnic Institute
- SERI: Solar Energy Research Institute
- SERDP: Environmental Research and Development Program
- SMES: superconducting magnetic energy storage
- UAW: United Auto Workers
- USG: United States government
- UT: University of Tennessee
- WCRE: World Council on Renewable Energy

Introduction

Many, if not most, readers will be largely unfamiliar with how the U.S. government ("Washington") works and policy is formulated and codified into law. That was me when I first arrived in DC in 1974 to work on Capitol Hill ("the Hill"), the name usually applied to the government's legislative center. I have always believed that my first and subsequent years on the Hill were an amazing period of education for me, for which I am grateful. Many people contributed to that education and they are mentioned in early numbered chapters of this book.

Capitol Hill

The U.S. government (USG) is an important and ongoing experiment on how people can govern themselves in a democratic fashion. In many ways it stands as an example to other nations around the world, in both positive and negative ways. Winston Churchill summed it up effectively when he stated: "Many forms of government have been tried, and will be tried in this world of sin and woe. No one pretends that democracy is perfect or all-wise. Indeed it has been said that democracy is the worst form of government except for all those other forms that have been tried from time to time." (House of Commons, 11 November 1947).

As we all learn in school, the USG has three independent branches. The legislative branch, Congress, has two parts, the House of Representatives ("the House") and the Senate. Both, and their surrounding buildings, are located on Capitol Hill. The executive branch consists of the President, the President's office, and a variety of federal departments and agencies (the "ministries"). The third branch is the Judiciary with its Chief and Associate Justices on the Supreme Court. This book discusses how energy-related laws are created under this system.

A word about the differing roles of the legislative and executive branches. The legislative branch, Congress, is responsible for enacting the nation's laws and appropriating the money necessary to operate the government. The executive branch is responsible for implementing and administering the public policies enacted by the legislative branch. This arrangement, which limits one branch from exercising the core functions of another, is referred to as a separation of powers. The intent is to prevent the concentration of power in any one branch and provide for so-called "checks and balances." In practice, the President prepares and presents a budget to Congress ("proposes") and Congress makes budget decisions ("disposes").

When one works on Capitol Hill, the first lesson one is often taught is the importance of jurisdiction—i.e., which committee of the House or Senate gets to consider a piece of proposed legislation ("bill") that is submitted by a member of that branch of Congress. These committees get the first bite-at-the-apple on that bill before it is brought to the floor of the House or Senate for broader review. Such committees have control of what comes out of committee after a vote by committee members and can even stop a bill from emerging at all. Control of such committees is in the hands of the majority party at the time and is reflected in more majority than minority committee members.

Once a bill is passed out of committee and goes to the floor, a set amount of time is usually allocated for full chamber discussion of the bill. One exception is a filibuster, defined by Wikipedia as "a dilatory or obstructive tactic used in the United States Senate to prevent a measure from being brought to a vote." A Senator may speak for as long as he or she wishes, on any topic, unless cloture (a 60-vote majority) is imposed and further discussion by the Senator is not allowed. Filibusters are not often used but have become more common in recent years.

Congressional approval means that a majority of the House (or Senate) has voted to accept the provisions of the bill. To move from Congressional approval to Presidential signature and codification into law a bill must be approved in identical fashion by both houses of Congress. This is not easy to achieve, given the broad range of views in a bicameral legislature of 535 members (435 House, 100 Senate). The second and third lessons I was taught when I first started working on Capitol Hill were the importance of establishing relationships and trust if one is to be successful on the Hill and the extremely small likelihood that any individual bill I would be working on will become law. I was told that a staff person has to derive his or her satisfaction from helping members gather and evaluate information, organize hearings, prepare draft legislation, respond to public requests, and generally help facilitate the legislative process. To put a fine point on this, in the 114th Congress, which began on January 3, 2015, and will last until January 3, 2017, as of late February 2016 4,570 bills had been introduced in the House and 2,548 in the Senate. As of that date, only 124 had been enacted into law and 7 were not enacted after a Presidential veto.

Assuming that a bill gets through the House and the Senate, the two forms of the bill are often not the same. This requires a meeting between selected members of both houses to iron out the differences in a "Conference." Both bills arrive at the Conference along with a "report" that accompanied each bill to the House or Senate floor and explains the provisions in the bill. While these reports have no official standing as legislation, they are important to eventual implementation of bills that are enacted into law—they reflect the intent of Congress. It is in the Conference where compromises are made and it can be a quick or a long and tedious process. Much of the negotiation is done at the staff level, in accord with member guidance. I worked on one bill where the differences were ironed out before House and Senate votes were taken and no Conference was required—the approved bill went directly to the President's desk. Another bill required several months of Conference (July through December) before agreement was reached—this bill is discussed in Chapter 3. It is typical in these Conferences to prepare printed side-by-sides, House provisions on one side and comparable but different Senate provisions on the other side. It is on the basis of these words that provisions are debated and final mutually agreeable language is decided. If this process is successful (not always), a final bill is prepared incorporating the agreed-to language and presented to both houses for final approval. This compromise bill is called the Conference Report.

One other word about the Congressional process. Ultimate decisions are made by elected members of Congress but staff play a major role given the many demands on a member's time and their lack of expertise on many of the issues that require their attention. Much of the negotiation and almost all of the drafting is done by staff, but accountability is high. A staffer going off on his or her own without authorization by the member is often grounds for firing.

This book will discuss my experiences working on the Hill and my subsequent experiences in the executive branch. Hopefully, this brief introduction will assist readers in better understanding what I am writing about in this series of recollections.

Chapter 1

The New England Years: How It All Got Started

I did not train for a government career in Washington, DC. In terms of energy as a profession, it all started for me on Saturday, December 20, 1969, when I attended a meeting at Amherst College as a favor to a friend and colleague. Up until that point, I had used energy like everyone else in traditional ways, but energy was part of the background and not of particular interest. That meeting changed my life and put me on a path that I'm still pursuing 47 years later. Here is how that change came about.

In 1969 I was a young physics professor at the University of Massachusetts in Amherst, MA, anticipating a traditional academic career in teaching and research (low-temperature solid-state physics). I had arrived at UMass a year earlier, after attaining my PhD at Brown University and serving as a post-doc at Brown for 19 months. I was working hard to set up my laboratory, teach my undergraduate physics courses, and attract high-quality graduate students to work with me in the lab.

In September of that year, a new member had joined the UMass Physics and Astronomy faculty, Dr. David Rittenhouse Inglis, who had just retired from Argonne National Laboratory at the age of 65. He was a distinguished theoretical physicist who had been part of the Manhattan Project at Los Alamos during World War II. He was

The U.S. Government & Renewable Energy: A Winding Road
Allan R. Hoffman
Copyright © 2016 Allan R. Hoffman
ISBN 978-981-4745-84-0 (Paperback), 978-981-4745-85-7 (eBook)
www.panstanford.com

also a founding member of the Federation of Atomic Scientists (now the Federation of American Scientists), and a leading voice in the effort to control proliferation of nuclear weapons. He was central to the successful effort to ban underground nuclear tests in the early 1960s.

David came from a long line of American scientists—one of his ancestors, David Rittenhouse, was a renowned astronomer and mathematician, and served as the first director of the U.S. Mint. When he arrived in Amherst, he was engaged in writing a book on energy and arms control issues, and started teaching a new course on those topics. This was the period when science courses were being designed specifically for non-science majors who had to meet a science requirement for graduation. David's course fit right into that requirement.

Somehow this young assistant professor and this senior full professor became fast friends. Thus, a few months later when he asked me to attend a Saturday morning meeting at Amherst College to which he had been invited but was unable to attend because of a conflict, I was pleased to say yes. He asked me to observe what happened at the meeting and then report back to him but never told me in advance what the meeting was about. It turned out to be a meeting of New England citizens concerned about plans to build a nuclear power plant in Vernon, Vermont, just 30 miles north of Amherst. I attended dutifully, listened carefully, and afterwards told David what I had heard. What came out of that meeting was a new environmental group, the New England Coalition on Nuclear Pollution (NECNP), dedicated to stopping the nuclear plant construction.

After filling David in on the details I also stated that while I was a trained physicist and understood the basics of nuclear fission, I realized that I knew almost nothing about the commercial application of nuclear energy and wanted to learn more. I had spent some time as an undergraduate getting familiar with nuclear fission and nuclear weapon technology, and had spent the summer of 1959 in New Mexico at Los Alamos National Laboratory, the home of the Manhattan Project. Even as a child of eight in 1945, when I heard the radio report in my parents' car about the dropping of an atomic bomb on Hiroshima and knew little of anything, I had a child's sense that this technology was a game changer. The decision to drop the bomb has always engaged my interest. Later, when I studied the physics of the fission process I would sometimes refer to it as "technologically sweet." I still feel that way from a physics point of view, but now have a much more nuanced view of nuclear fission as a weapon of war and as an energy supply option.

David knew a lot about nuclear fission from his physics studies, his time at Los Alamos, and his subsequent several decades of work on nuclear issues at Argonne. Nevertheless, he also felt that he had limited knowledge on how nuclear power plants operated in the real world, and he and I quickly agreed to organize a physics faculty discussion group to discuss nuclear issues at lunch once a week on Wednesdays at the UMass faculty club. We ended up with six members, five from UMass and one, Bob Romer, from Amherst College. We met regularly for close to a year, benefitting mostly from David's knowledge, but also benefitting from weekly research done and reported by the other members. At the end of that year, I went to David and said that I found the pace of learning too slow. As a young faculty member, I had come to understand that one important way to learn something is to teach it and try to stay ahead of your students, and so I asked David if I could teach a second section of his course in the upcoming semester. He quickly agreed, and I was soon teaching about 100 non-science students the basics of nuclear weapons and nuclear power plants and eventually inherited the course from David when he stopped classroom teaching.

The course was a lecture course, in the style of the times, with occasional guest speakers. As you might expect, in the course of reading the literature and preparing lectures, I was learning more than any of my students, and by my second semester of teaching this course, I was quite comfortable with the subject matter. During these months NECNP also went public, and both David and I followed its activities closely. While neither of us was anti-nuclear in any sense, we both realized that nuclear power had both its upsides and its downsides and was a topic worthy of extended public discussion. This led us to attend an evening meeting of NECNP in Brattleboro, VT, not too far from Amherst. It was a meeting that had major ramifications for me.

As David and I listened to and observed the NECNP discussion that night, we both realized that NECNP had no technical backup. Its members were dedicated lay people but there was no one with an engineering or science background. In order to ensure that any press releases coming out of NECNP would be technically sound, we both volunteered to serve as technical advisors to NECNP and reviewers of their public statements. When this became known to the anti-nuclear community in New England, we began to receive invitations to speak about and debate nuclear power issues, and I ended up accepting most of the invitations as David was a reluctant public speaker.

About this time nuclear power first began to be noticed as a public issue, stimulated in part by increasing awareness of the military and civilian nuclear waste problem and the Nixon Administration's unsuccessful effort to promote Lyons, Kansas, and its salt beds as the solution to this problem. The rationale being put forward was that salt beds were inherently dry and would quickly encase any nuclear wastes that were buried in its depths. This argument fell apart quickly when it was pointed out that salt beds often overlay oil deposits and that several exploratory oil wells had already been drilled into the beds, providing a path for water to reach objects buried in the salt. Today the world is still wrestling with the need to identify a safe way of isolating these wastes for long periods of time.

In my invited talks, starting in 1970, I attempted to present the pros and cons of nuclear power, often to audiences that were more anti- than pro-nuclear. In the three debates I participated in, I was usually cast as the one pointing up nuclear power's problems while my debate counterparts were making the case for nuclear power. These counterparts included a physicist from Brookhaven National Laboratory who famously said that "it was more dangerous to drink a cup of coffee than to live near a nuclear power plant," a member of MIT's nuclear engineering faculty who spoke in a balanced way about low-risk, high-consequence events, and a vice president of Northeast Utilities whom I debated on public television in Boston—Northeast Utilities was heavily into nuclear power. It was also during this period that I started sending letters to Walter Sullivan, the Science Editor at the *New York Times*, encouraging him to use his public platform to encourage a public discussion of nuclear power issues (both David and I felt strongly that such a discussion was needed). He ignored the two letters I sent to him, and out of frustration David and I met to discuss how to get this discussion started. Our answer was to get Ralph Nader involved—Ralph was then becoming famous for raising issues about automobile safety and was well known to the public. We agreed to invite Ralph and his PIRG (Public Interest Research Group) directors to Amherst for a weekend of briefings on nuclear power issues, to end with a request to go forth and raise the issues in a highly visible way. Ralph accepted, the weekend happened, and, to some extent, the rest is history.

Nuclear power got a lot of public attention in the 1970s, especially after the Arab oil embargo of 1973–1974 and utilities began to build nuclear power plants as a hedge against importing oil from other countries. Oil was then an important fuel for generating electricity. In fact, at one point in the mid-1970s, a common mantra expressed by the U.S. nuclear industry and some politicians was "a doubling every decade" of the number of U.S. nuclear power plants for the next three decades. This view was supported not only by nuclear power advocates but also by those who believed that economic growth as measured by Gross National Product (GNP) tracked one-to-one with energy consumption. They derided the arguments of those who argued that growth could be achieved along with improved energy efficiency and that the direct historical linkage between energy consumption and GNP growth was no longer valid. It took many years, but eventually it was shown that GNP and energy consumption are not tied closely together.

I finally met the *New York Times*' Science Editor at an American Physical Society meeting in New York, after nuclear power had begun to attract some attention. I introduced myself, kindly confronted him about his failure to even acknowledge my letters, and he graciously admitted that he had been wrong not to respond.

Back in Amherst, after several speaking engagements and interacting with and answering questions from the audiences, I realized that if people were going to be against nuclear power for whatever reason, they would have to be in favor of growing the use of some other energy source or sources since the world's population was growing and more energy would be needed to meet people's needs. This was my initial path to thinking about renewable energy. I began to educate myself on the subject and quickly became an advocate. I met with other young enthusiasts in the New England area who were looking at solar and wind energy projects, and I even proposed a solar-powered walkway for downtown Amherst to demonstrate the possibilities, which attracted some attention from the local press.

About this time I learned about a new program sponsored by several technical societies, including my own, the American Physical Society (APS), to bring PhD-level scientists and engineers to Washington, DC, to work with Congress on technical issues. Congress' agenda was becoming increasingly technical and Congress lacked the needed technical staff capabilities. It was called the Congressional Fellowship Program, had brought seven scientists and engineers to DC in 1973 for one-year fellowships supported by the societies, and was now advertising for its second class that would come to Washington in September 1974. This got my attention as my interest in energy issues was growing and I was curious as to how these issues were being addressed by Congress. I decided to apply and was a finalist in two fellowship programs, the one offered by the APS and the one offered by the American Association for the Advancement of Science (AAAS). Interviews were scheduled for March, the APS interview came up first, and my interview was scheduled for a day during the APS' Spring Meeting in DC when I could not attend—I was giving a speech on nuclear power that evening in Connecticut (CT). I called APS, offered to be in DC first thing the next morning by sleeping at a motel near Bradley Airport in CT so I could catch the first plane in the morning to DC. They accepted this offer, and I showed up just in time to learn that APS had awarded two fellowships the previous day. This was important information as I had been told that the APS Council had authorized up to two fellowships per year. Nevertheless, they went ahead with my interview, which involved a presentation to a panel of physicists on a topic of my choosing, to simulate a briefing for the members of Congress. I chose nuclear power to talk about, given my familiarity with the subject and realized quickly after meeting the panel that everyone in that room was probably a strong advocate for nuclear. As far as I could tell, it went well and shortly after the interview I was informed by Dr. Mary Shoaf, the head of the APS Fellowship Program, that I was also to be awarded a fellowship, the only year in the history of the Program that more than two fellowships have been awarded. I was very pleased and began planning with my family (wife, two kids ages 9 and 12) for what I anticipated would be a one-year assignment and slight but interesting diversion from my academic career. My attitude toward the impending fellowship experience was largely positive but colored by concern about living in Washington, given all the media stories about Congressional shenanigans, an often unpleasant summer climate, and a high level of crime. Nevertheless, I figured I could do anything for a year, and it would be a relatively unique experience.

Chapter 2

Introduction to Capitol Hill

In mid-August my family and I traveled from Amherst to the DC area in a rented truck and our family car, arriving two weeks after President Nixon resigned his office. I was not unhappy to see Nixon gone, having long distrusted the man and the people around him. His departure before formal impeachment also simplified my fellowship year a great deal—before his resignation I was concerned that Congress would spend most of the following Congressional session in impeachment proceedings and my chance to get involved in policy issues was going to be limited if not eliminated. His departure from DC allowed Congress to return to its normal activities and begin to address the impacts of the Oil Embargo that had been imposed by the Organization of Petroleum Exporting Countries (OPEC) in the preceding months. It had been imposed in response to U.S. assistance to Israel in the 1973 Middle East War and quickly reduced U.S. imports of oil from OPEC nations by approximately 30%. This created a bit of panic in the United States, which had not experienced anything like this before, caused gasoline prices to rise (from about 20 cents to 39 cents a gallon), and saw the imposition of alternate day access to gasoline stations based on your license plate number. It even led to fistfights when anxious and frustrated drivers tried to cut ahead into long lines at the pumps and those in line objected.

The U.S. Government & Renewable Energy: A Winding Road
Allan R. Hoffman
Copyright © 2016 Allan R. Hoffman
ISBN 978-981-4745-84-0 (Paperback), 978-981-4745-85-7 (eBook)
www.panstanford.com

It was a real wakeup call for the country, which had been the world's largest oil producer for many years and for which energy had not been an area of concern. In my Fellowship year, the second year of the program, there were 12 Fellows, supported by various societies—AAAS, APS, ACS (American Chemical Society), ASME (American Society of Mechanical Engineers), and a few others. All Fellows underwent a month of orientation organized by the AAAS, to get us ready for what we would experience on Capitol Hill. This included reading books on Congressional process, briefings from Hill Staff, a briefing from a senior official of OMB (the Executive Branch Office of Management and Budget), and many hours spent in interviews with Congressional offices and committees. Most of us were not given assignments before we arrived—this was to be a process where Fellows would find their own assignments, which could have been anywhere on the Hill. A few of the Fellows were being assigned to the new Office of Technology Assessment (OTA), but their specific assignments within OTA were still to be negotiated.

It is important to note that in 1974 the Fellowship Program was still brand new to both the professional societies and the Hill and there were many uncertainties. Despite an apparently successful first year of the Program, it was still unclear whether the Hill would be open to taking scientists and engineers on their staffs when the majority of staff at the time had legal, economic, or political science backgrounds. It was also not clear how valuable the scientists and engineers would find the Hill experience, and whether in a one-year assignment they could make a difference. Ground rules were still being set on independence of the Fellows from their professional societies, and levels of trust were still to be established between the political set and the technically oriented new kids on the block.

My job interviews, which reflected both my interests and expressions of interest from offices willing to entertain the thought of a Fellow on their staffs, involved both individual Congressional offices and Congressional committees. My first interview was with the staff in the office of Senator William Proxmire (D-WI) who was well known for his skepticism about certain government-funded R&D projects (he invented the "Golden Fleece" award). The interview went well until the Staff Director asked me: "Why would a scientist want to get his hands dirty with a dirty Congress?" To me this reflected a serious misunderstanding of why the Fellowship Program existed and reflected the biases we were warned to expect. I decided this was not an office I wanted to be a part of.

On a more positive note I interviewed with senior staff in the office of Senator Chuck Percy (R-IL) and they offered me a job with a focus on foreign policy. I also interviewed with the Legislative Director of the House Interior Committee chaired by Mo Udall (D-AZ), and who was a Fellow from the Program's first year, and with the Staff Director and Legislative Counsel of the Senate Commerce Committee chaired by Warren Magnuson (D-WA). Both Committees offered me staff positions, putting me into a quandary with three attractive offers. I finally resolved it by deciding I wanted a Committee rather than an individual office assignment, to allow me exposure to a wider range of issues, and by going to the Library of Congress one morning and reading the *Washington Post* while my brain considered my decision. After a bit of this cogitation I decided to accept the Commerce Committee offer, walked over to their office in the Senate Dirksen Office Building, was designated as their Staff Scientist and assigned to the Science Subcommittee chaired by Senator Phil Hart of Michigan. I could not have been more pleased—Senator Hart was known as the conscience of the Senate and was so highly regarded by his colleagues that they eventually named the new Senate Office Building after him.

Senator Phil Hart

While Senator Hart chaired the Subcommittee, my specific assignments were to come from Senator Magnuson's office, headed by Staff Director Mike Pertschuk and his close associate and Staff Counsel Lynn Sutcliffe. During my time on the Hill I worked most closely with Lynn, a brilliant strategist and dedicated public servant, from whom I learned a great deal. I also worked closely with three other staffers, Mike Brownlee, Dan Jaffe, and a more senior staff member, David Freeman, who joined the Committee staff on the same day that I did, October 1, 1974. Mike took me under his wing that first day and began to teach me the ropes about operating on the Hill, educating me about the importance of Committee jurisdictions (which Committee gets to review and act on submitted legislation), the very small chance that any legislation I would work on would get enacted into law, and the importance of relationships and trust in getting anything done on the Hill. He and I ended up working on different bills, but I have always been extremely grateful to Mike for his wise counsel that day and on many days thereafter. Dan, a lawyer, also worked mostly on bills I was not involved with, but he was one of the people I shared an office with (Mike was the third) and we became good friends. Dave was older than the three of us, already had a well-established reputation as a visionary on energy issues, and even had his own office next to ours. He eventually went on to head the Tennessee Valley Authority in the Carter Administration and subsequently major utility organizations in Texas, California, and New York. He remains a friend to this day.

I was the only scientist on the Commerce Committee staff. When the Congressional Fellowship Program started in 1973 there were only a handful of technical staff on the Hill, and in 1974 that number was still very small compared to the number of highly trained scientists and engineers in the Executive Branch. The purpose of the Program was to provide assistance to Congress as it addressed increasingly technical issues. The program proved so successful that it is still going today, more than 40 years later, and has been extended to Executive Branch departments and agencies. In recent years more than 200 Fellows have been brought to DC annually, and about 35 are still assigned annually to Congressional offices.

My first assignment was in response to the Oil Embargo: figure out how to set up a rationing system for gasoline in the United States since most oil in the United States was used to support transportation. It required me to call people who had been involved in rationing during World War II, under the Office of Price Administration. Many of these people were still alive and it was a rather enjoyable assignment. The then-existent Federal Energy Administration (FEA) in the Executive Branch also got involved and actually printed gas rationing coupons for use when and if needed. I heard many years later that they had been stored in a warehouse in the DC area, but in truth I still don't know what became of them.

After about two weeks of such calling and preliminary planning, it became clear to my masters on the Hill that a gasoline rationing system was not politically feasible and the idea was dropped. This led immediately to my second assignment: figure out how to reduce U.S. oil imports. This was not an easy assignment for someone who had been on the Hill for just a few weeks. It also meant that my first involvement with energy issues in DC would be on the energy efficiency side, even though the future arc of my energy career would focus on renewable energy.

Chapter 3

OPEC Oil Embargo and Corporate Average Fuel Economy

In thinking about the oil import issue, with the aid of Bob Hemphill and two of his staff at FEA, it quickly became apparent that one had to think about impacting the biggest consumer of oil in the United States, consumption in transportation. Gasoline used in cars accounted for about two thirds of total oil use.

1974 Cadillac Coup de Ville

The U.S. Government & Renewable Energy: A Winding Road
Allan R. Hoffman
Copyright © 2016 Allan R. Hoffman
ISBN 978-981-4745-84-0 (Paperback), 978-981-4745-85-7 (eBook)
www.panstanford.com

Gasoline consumption in cars is easy to understand as it depends on just three numbers—the number of cars on the road (N), the annual vehicle miles traveled by the average car (known as VMT), and the average number of gallons used per VMT. This latter factor, measured in gallons per mile, is the inverse of what we routinely refer to as a car's fuel economy (FE) in miles per gallon (mpg).

$$\text{Total gasoline consumption in cars} = (N) \times (\text{VMT}) \times \frac{1}{(\text{FE})}$$

It was clear that N was an untouchable number—reducing N would be considered "unAmerican" and possibly lead to a new American revolution—and reducing VMTs would require an increase in the price of gasoline, a prerogative of Congress. At that time gasoline prices had doubled from their very low levels before the OPEC Oil Embargo and nobody was in the mood for another increase. Nevertheless, legislation had been introduced in the House of Representatives to raise the federal gasoline tax by 3.5 cents per gallon, and in early February 1975, shortly after the new 182nd Congress had convened, I walked over to the House to observe the vote from the House Gallery (Note: the Senate allows Congressional staff on the floor of its chamber, but not the House). The members were clearly concerned about the threat to our energy supplies posed by the Embargo, but with a large country requiring lots of automobile travel, especially in the more rural sections of the country, there was concern about the political ramifications of making it more expensive for people to drive. This is still true today. As I sat in the Gallery, I watched the House vote down the 3.5 cent increase, as well as subsequent motions to reduce the tax increase to 3.0, 2.5, 2.0, 1.5, 1.0, and 0.5 cents per gallon. All were voted down. As I walked back from the House late that afternoon to my office on the Senate side of the Capitol, I realized that if Congress was unwilling to increase the cost of gasoline, and reducing N was a non-starter, then the only factor subject to being changed was the average fuel economy of the American automobile fleet, which at that time was about 14 mpg. This was the origin of what are now known as the Corporate Average Fuel Economy (CAFE) standards.

The next day I reviewed what had been done about fuel economy issues in the previous Congress, largely by Dr. Barry Hyman, a professor of mechanical engineering at George Washington University and the first ASME Congressional Fellow in 1973, and began to educate myself about options for improving automobile fuel economy. I did this by reading the literature (and please remember that this was in the days before personal computers and search engines) and talking with experts in the industry, the government, and in academe who were familiar with automobile technology. As a Congressional staffer I also had access to the excellent research capabilities of the staff at the Library of Congress, and easy access to lots of people when I called and used my Congressional staff title. It was an intriguing education—I knew little about automobile fuel economy beforehand, having been trained as a low-temperature solid-state physicist—and soon I was at the point of putting some thoughts on paper and preparing a draft piece of legislation. It was the first bill I ever wrote, and the most important. Also, I was now working closely with Senator Fritz Hollings (D-SC), a member of the Subcommittee whom I came to respect highly and with whom it was a pleasure to work. When I eventually left my position on the Senate staff a few years later, I spent an hour alone with the Senator in his office saying our goodbyes.

When the first draft was ready for presentation to the full Commerce Committee, Senator Magnuson called a meeting of its members in the Committee's conference room, and staff were invited. The draft called for a 10-year increase in fuel economy, averaged over new automobile fleets, to 28 mpg. I had concluded from my studies and discussions that a doubling of national average fuel economy over 10 years was a stretch but doable goal, and Sen. Hollings agreed. It was a mandated performance standard, not a prescriptive standard, and the auto industry was to be left on its own to decide how to meet the standard. They were to be given the end point to be reached 10 model years later, with intermediate standards along the way to monitor progress.

Sen. Hollings made the presentation to the committee and led the discussion—staff were to be seen but generally not heard—and received negative feedback from only one Senator, Sen. Robert Griffin (R-MI). It was no surprise that Sen. Griffin might raise some objections to any controls on the auto industry—after all he was from Michigan, the home of the industry, and as a Republican he had strong support from the automobile companies.

The discussion between the two Senators went along smoothly until Sen. Griffin pointed out that the proposed legislation was quite different from draft legislation the members had seen in the previous Congress just a few months earlier. In his response Sen. Hollings brought the room to complete silence by stating "consistency is the hobgoblin of feeble minds," a slight misquote of what Ralph Waldo Emerson actually said: "A foolish consistency is the hobgoblin of little minds, adored by little statesmen and philosophers and divines." Members were stunned that one Senator would speak to another Senator in that manner, especially with staff in the room, and the meeting came to an abrupt end shortly thereafter.

With signoff by Sen. Magnuson and the Democratic majority of the Committee, including Sen. Hart, who was also from Michigan, the next few months were spent in setting up Congressional hearings, improving the draft, and bringing the draft legislation to the House side of the Hill where legislative action was also required if the bill were to eventually be sent to the President for signature and enactment. The House was also in Democratic hands at that time, and the bill was brought to the House Committee on Commerce and Transportation chaired by Rep. John Dingell (D-MI). For the rest of 1975 my principal staff companions were Lynn on the Senate side and Bob Nordhaus, Charlie Curtis, and Shelly Fidler on the House side. Charley and Bob were Rep. Dingell's chief legislative assistants on the House Committee and Shelley played the same role for Rep. Phil Sharp (D-IN) who was a member of the Committee.

I arranged several hearings on the proposed legislation for the Senate Committee, and as expected there was a wide range of opinions on the proposed legislation. Those who testified from the auto industry strongly opposed any legislation calling for fuel economy standards, arguing that they were already required to reduce emissions by the Clean Air Act and that they could not at the same time increase fuel economy. This view was strongly opposed by others familiar with the industry, including the head of the UAW (United Auto Workers). He informed the Committee that he had been urging the industry for several years to increase fuel economy, to respond to increasing market penetration by more fuel efficient Japanese cars, but that the industry wouldn't do it because the profit margin on big cars was bigger than that on smaller cars—you could charge so much more for a bigger car when the only additional production cost was for a bit more steel. This view was corroborated by a former industry subcontractor who stated: "The only way to get the industry to increase fuel economy is to force them." The industry also argued that the proposed legislation would restrict the number of cars that would be available to pull recreational vehicles. This argument was presented to Shelley by General Motors (GM) along with a briefing book in support of their argument. Shelley called me, asked for help, and forwarded the GM briefing book to my office. Calculations quickly revealed that this argument was unsupportable, I reported this to Shelley, and the issue went away.

Another issue that arose was how to deal with luxury car fleets that were unlikely to meet the standards. Further calculations revealed that the amount of gasoline likely to be used by these luxury cars was small, and I recommended that we let the luxury car purchasers pay the proposed civil penalty for non-compliance and leave it at that, recognizing that we couldn't fix all the problems in one bill. Of course, we were subjected to considerable lobbying on all sides of the fuel economy issue, including one day when Lynn and I met with supporters of the legislation in the morning, from the Nader organization Public Citizen, and industry opponents of the legislation in the afternoon. Our end-of-day private conclusion was that we must be doing something right.

The House version of the legislation, which set the 1985 new car fleet fuel economy standard at 27.5 rather than the Senate's 28 mpg, reached the House floor in June, while the Senate version was scheduled for debate in July. I sat in the House Gallery the day of the debate, next to a colleague who worked for the National Automobile Dealers' Association (NADA). He was someone I had become friendly with during the intervening months. To my great surprise he was rooting for the bill to pass by a 4-to-1 margin. I remember turning to him and saying: "What's wrong with you? You work for the automobile industry and they are strongly opposed to this legislation. Why are you rooting for it to pass?" His answer: "I've told the industry that this bill is going to pass and they don't believe it. They don't think that Congress has the balls to pass it." Well, he was right and they were wrong—the bill passed by a margin of three and a half to one.

The bill reached the Senate floor about a month later, and I assisted Sen. Hollings while he managed the 6-hour floor debate. He had prepared himself well, not only rereading the bill the night before, but also reading the full explanatory report accompanying the bill to the floor, which I had drafted and which he subsequently sent to every automobile dealer in the state of South Carolina. It was also my first time on the Senate floor. The final vote in the Senate that day in favor of the CAFE legislation was 63 to 21.

The next several months of 1975 were spent in conference with the House, to resolve the differences between the two versions of the Energy Policy and Conservation Act (EPCA) that had emerged from the floor actions. At the staff level, this effort on the Senate side was led by Lynn, while Charlie and Bob led the House effort. CAFE was only one of several titles that were proposed as new energy policies for the United States, and it took until December 1975 to resolve the differences and bring a conference report (the bill agreed to in conference by the House and Senate negotiators) to a vote in both Houses. It was quickly passed and signed into law by President Ford just before Christmas. The signing was followed by a brief celebration among House and Senate staff members who had been most involved in the negotiations, at the Hawk and Dove watering hole/restaurant popular with Hill people. We were accompanied by Rep. Dingell, who had been a consistent supporter of the legislation. He even offered to pay for the drinks, a kind offer that was appreciated but refused. (Note: These were not easy decisions for either President Ford or Rep. Dingell, both of whom were from Michigan. It was an example of politicians putting the interests of the country first.)

One final conference anecdote about why the final 1985 standard was set at 27.5 mpg. This number derives from the House version of the legislation. Midway through the conference deliberations Lynn approached me and asked what our position should be: 28 or 27.5? A few quick calculations revealed that the House method of calculating the average was slightly more stringent than the Senate method, and would lead to slightly greater fuel savings. My advice to Lynn, which he accepted, was to accede to the House position, the House would be pleased that we'd accepted their version of the legislation and gain us some bargaining leverage in other conference negotiations, and the country would benefit from a slightly more stringent standard. Hence, 27.5.

Several months after CAFE became law, an oversight hearing on its implementation was held by Senator Adlai Stevenson, Jr., Chairman of the Senate Commerce Committee's Science Subcommittee. I staffed the hearing. It was a difficult hearing, during which senior representatives of the U.S. automobile industry continued to insist that they couldn't achieve the mandated 1985 and intermediate standards while reducing exhaust emissions. Having been told by others that the industry would resist strongly in its testimony but ultimately not violate a law of the United States, I quietly passed a note to the Senator asking him to put each of the executives on the record: Will your company meet the standards? They all testified yes.

A final piece of CAFE history: about a year after the legislation was signed into law, I ran into the chief lobbyist for one of the automobile companies in the U.S. Capitol. He pulled me aside, told me he would never say this publicly, and expressed his opinion that the legislation had "saved his industry"—this was a time of significant loss of market share to Japanese auto companies. That may not have been true and I'm sure that many in the industry would have strongly disagreed with his statement. Nevertheless, those of us who worked on CAFE take pride in helping the country move forward after the Oil Embargo. The legislation achieved its goal of improving new automobile fleet fuel efficiency, but, unfortunately, by reducing the cost of driving it stimulated VMT increases, which, in the absence of an increase in the cost of gasoline, partially offset the possible fuel savings. This was a lesson for the future.

The *New York Times*, in a June 19, 2007, editorial entitled "Crunch Time on Energy," stated: "The most effective energy efficiency policy ever adopted by the federal government is the Corporate Average Fuel Economy requirement of 1975." This was at a time when Congress finally made minor adjustments to the standards after doing nothing for 32 years. Many years had passed since CAFE was enacted and implemented, a period during which oil imports and oil prices had increased dramatically, and climate change had been recognized as a serious global challenge. A more recent response has been the agreement between the Obama Administration and the auto industry to increase the CAFE standard to 55.4 mpg in 2025, an important further step in establishing energy efficiency as the cornerstone of U.S. energy policy.

While this chapter has focused on CAFE, I also was involved with several other issues for the Commerce Committee during my time on the Hill. In 1977 the Senate was reorganized and the Commerce Committee absorbed the responsibilities of the Committee on Space and was renamed the Commerce, Science and Transportation Committee. As I was still the only scientist on the staff (a second Congressional Fellow was added to the staff shortly before I left the Senate) I continued to handle "technical stuff," including oversight over NASA and the National Bureau of Standards (which was renamed the National Institute of Standards and Technology, NIST) and energy efficiency legislation. I also drafted legislation that returned the office of Science Advisor to the White House—President Nixon had kicked the Advisor out of the White House in 1973 when he was displeased with the advice he was receiving from the Advisor's office on the supersonic transport (SST) R&D program. He was for the SST; the Science Advisor recommended against it. The President then added the Advisor's responsibilities to those of the Director of the National Science Foundation (NSF).

I was also involved with other energy efficiency legislation (auto R&D and industrial process R&D), including two overrides of Presidential vetoes (we failed on an attempted override of a third) and passage of an energy efficiency bill introduced by Sen. Ted Kennedy and assigned to the Commerce Committee. It sticks in my memory because on the day that the bill was due for debate on the Senate floor, Sen. Kennedy could not attend and his close friend Sen. Hollings agreed to handle the floor debate for him. I was on the Senate floor with Sen. Hollings and there was a lively debate with another Senator who was not thrilled with the proposed legislation. It got pretty heated and at some point the other Senator stomped angrily off the floor. I went over to the other Senator's staffer and asked what had happened. He told me that his Senator was upset with the tone of Sen. Hollings' comments and vowed "never to debate Fritz Hollings on the floor of the U.S. Senate again."

I was also asked to handle the issue of the Soviet Union's bombardment of the U.S. Embassy in Moscow with microwave radiation, presumably in an attempt to monitor and decipher window vibrations arising from discussions inside (interesting physics problem). In addition, I organized and staffed hearings on nominations for the heads of several Executive Branch science agencies: Administrator of NASA (Bob Frosch) at a time when the Shuttle was being designed and tested, Science Advisor to the President (Frank Press), and Undersecretary of Commerce (Jordan Baruch). All in all, in the three and a half years I spent on the Hill, I had a broad exposure to many issues, a unique education, and a few important impacts. Another result was my family's decision in 1978 to stay in the DC area and not return to Massachusetts. The basis for this decision is discussed in Chapter 4.

Chapter 4

First Tour of Duty in the Executive Branch and the DPR

The end of 1976 saw the election of Jimmy Carter as President of the United States and extensive discussion of the need for a new federal agency dedicated to addressing national energy issues. Thus was created the U.S. Department of Energy on October 1, 1977, and its absorption of the FEA, the Energy Research and Development Administration (ERDA), the Research Applied to National Needs (RANN) program of the NSF, and selected parts of the Department of the Interior (DOI). I was asked by the Carter Transition Team to comment on several energy-related issues, including management issues of a new department, and the advisability of continued funding of the federal government's fusion energy R&D program. On the former I recommended a Deputy Secretary focused on management issues because departmental Secretaries often devote too little time to management given their other responsibilities. On the latter I recommended continued funding based on the long-term potential of fusion energy, despite its many technical barriers. If successfully harnessed it is essentially an unlimited energy source with fewer and shorter term radiation waste disposal problems than nuclear fission. I was unsuccessful with the first recommendation and successful with the second.

The U.S. Government & Renewable Energy: A Winding Road
Allan R. Hoffman
Copyright © 2016 Allan R. Hoffman
ISBN 978-981-4745-84-0 (Paperback), 978-981-4745-85-7 (eBook)
www.panstanford.com

And then my life changed again. Early in 1977, while still serving as Staff Scientist for the Senate Commerce Committee, I was invited to join the Carter Administration as a political appointee in the Department of Transportation (DOT), with responsibility for the CAFE program which had been enacted into law at the end of 1975 and assigned to DOT. I interviewed with Joan Claybrook, Ralph Nader's close associate, who had been appointed head of DOT's National Highway Traffic and Safety Administration (NHTSA). It was an interesting opportunity, given my intimate association with CAFE, but I finally decided not to accept the position and limit my involvement in other interesting issues.

The next opportunity came quickly when I was then invited to meet with Al Alm, head of the new DOE's Office of Policy. Previously, Al had been a senior official at CEQ and EPA, and in the words of President Bill Clinton: "Al was one of the true heroes of the environmental movement in America."

Al Alm

He invited me to join his office as head of his nuclear policy operation. I responded that it was a position I was qualified to fill but that he might want to consider that I had been involved as an unofficial but widely known advisor to a citizens' group in New England that was opposed to nuclear power, and I wasn't sure he would want the political problems that might possibly arise from that connection. His immediate response was: "OK, what do you want?" I said renewable energy, was appointed head of his Office of Advanced Energy Systems, which had responsibility for renewable energy policy, and officially joined DOE as a political appointee on April 3, 1978. What I didn't appreciate at that time was how busy I would be in the next few months.

When I arrived on April 3, just a few months after DOE had been established, office assignments were still being figured out in the Department's headquarters, the Forrestal Building. It had been named years before after a former Secretary of Defense when the building was part of the Department of Defense (DoD). The new department had inherited it—a building I began to call "the concrete mausoleum" because of its drab concrete corridors—and things were a bit disorganized in the early days. I remember asking some colleagues who were veterans of the Executive Branch how long it takes for a new department to get its act together and the consensus was about 10 years. As a result my first office assignment was not in Forrestal but in the Old Post Office Building in downtown DC (soon to be a Trump luxury hotel), a magnificent building with beautiful staircases and mahogany-paneled offices from its early 20th century days as the home of the U.S. Postal Service. Joining me in these luxurious offices was one of two staff I inherited when I joined DOE, Ron White, an economist who had been working on regulatory issues. Unbeknownst to us, our stay in the Old Post Office Building was to be short-lived, for on May 3 President Carter traveled to Golden, CO, to dedicate the new Solar Energy Research Institute (SERI) and announced a new study, "A Domestic Policy Review of Solar Energy." A short while later that day, I was called into Al Alm's office and informed that I would be leading that multiple-agency study as DOE's senior representative to the study task force. Thus, one month after arriving at DOE I got what I had asked for in spades.

President Jimmy Carter

The first issues were getting an office in Forrestal, organizing a staff (in addition to Ron I inherited only one other staffer, Billy Owens), and organizing the efforts of the many detailees from the 29 other federal departments and agencies that would be contributing to the study. Getting to Forrestal quickly was not easy—DOE couldn't provide staff and trucks to move our offices for several days—and so Ron and I used our own cars to move our stuff into our new assigned space on the Forrestal's G level. It was a large open space, chosen to accommodate the expected large number of detailees and well-equipped with asbestos ceilings, water leaks, cockroaches, and the occasional mouse. In the early weeks of May it was outfitted with all the required desks in an open configuration with one exception, four walls surrounding a small space in the middle that served as my office.

Staffing up proved to be difficult. I knew I needed about a dozen DOE staff and hiring had to go through an antiquated personnel system. Positions were advertised widely both within DOE and externally, and I spent a lot of my time in those first few weeks, along with Ron, interviewing candidates. Unfortunately, all recommended hires and their salaries had to be approved by reviewers in the DOE Personnel Office and my assigned reviewer, while pleasant, had a drinking problem and was often not in the office. This required "pushing the envelope" in a few cases to get the needed staff on board quickly, but within a month I had hired the people I wanted. All were assigned to desks in the G-level space which I labeled "the bullpen."

The study was initiated formally by a May 16 memorandum from Stuart Eizenstat, Director of President Carter's Domestic Policy staff, and its scope was defined as follows: "A thorough review of the current Federal solar programs to determine whether they, taken as a whole, represent an optimal program for bringing solar technologies into widespread commercial use on an accelerated timetable; A sound analysis of the contribution which solar energy can make to U.S. and international energy demand, both in the short and longer term; Recommendations for an overall solar strategy to pull together Federal, State and private efforts to accelerate the use of solar technologies."

In response to this memorandum, a Cabinet-level interagency Solar Energy Policy Committee under the chairmanship of the

Secretary of Energy was formed to conduct the review. At the working level over 100 senior level detailees (175 at its peak) representing 30 executive departments and agencies were active participants starting in early June. (Note: the term "solar energy" in this memorandum referred to the full spectrum of renewable energy technologies that are direct or indirect forms of solar energy and so included solar photovoltaics, concentrating solar power, solar heating, wind power, biomass, hydropower, and various ocean energy technologies (ocean thermal energy conversion/OTEC, wave energy, and ocean current energy). Thus, a more accurate name for the study might have been "A Domestic Policy Review of Renewable Energy." In any event the study soon came to be called the DPR.)

An early issue was whether to include geothermal energy as part of the study. Geothermal energy is not solar-dependent either directly or indirectly and derives from radioactive decay in the earth's core. After listening to opinions from my staff on this issue, I decided to include geothermal energy, given its large energy potential and essentially renewable origins.

At my insistence the review was conducted with significant public participation. Twelve regional public forums were convened throughout the nation during June and July to receive public comments and recommendations on the development of national renewable energy policy. The response of the public was impressive and reflected the growing support for renewable energy that had been identified by several opinion polls. Several thousand people attended the meetings and over 2,000 individuals and organizations submitted oral or written comments.

In addition, briefings were given to members of the DPR staff by representatives of renewable energy advocacy groups, small businesses, state and local governments, public interest and consumer groups, utilities, renewable energy equipment manufacturers, and other parts of the energy industry. This public input was an important part of the DPR process.

In large part, themes reflected in the public comments were consistent with the eventual findings of the DPR, which was delivered to the White House on December 6, 1978. These themes were also consistent with the premises of the National Energy Plan that had been prepared and released just prior to the announcement of the DPR. These premises included an emphasis on conservation (today referred to as energy efficiency) as a cornerstone of national energy policy, awareness that energy prices should generally reflect the true replacement cost of energy, and recognition of the need to prepare for an orderly transition to an economy based on renewable energy resources. The public forum comments also reflected a deep concern that the poor and the elderly have access to affordable energy.

The next six months were rather intense, starting with the fact that the other 29 departments and agencies didn't trust the 30th, DOE, because of some recent history. The National Energy Plan was also a multi-agency effort chaired by the DOE. The story I was told by non-DOE staff was that the DOE, at the last minute, had pulled out a draft it had prepared on its own and submitted it as the multi-agency report. As a result I inherited a serious problem of distrust and spent much of the DPR's first month building relationships with the non-DOE detailees to reestablish that trust.

Over the next few months literature was reviewed, scenarios were considered, numbers were crunched, and report chapters were outlined, drafted and redrafted. Lots of pressure was focused on me and my DOE staff as we had responsibility for preparing the final draft for review by the cabinet-level committee before submission to the President. Let me give two examples of such pressure: (1) Gus Speth, head of the Council on Environmental Quality (CEQ), and his Deputy, Jim McKenzie, were in my office often because they knew me hardly at all (I had met Jim briefly in Massachusetts shortly before we both came to DC) and were concerned that I wouldn't be sufficiently friendly to renewable energy; and (2) I began to find notes on my desk from someone named Amory Lovins, whom I didn't know at the time, encouraging my support for renewable energy. The notes were apparently put on my desk by one of my staff, Dick Holt, who was a friend of Amory's and was connected to Amory's growing network. What I didn't know until recently was that Dick had invited Amory to visit Forrestal and asked Ron to show him around, obviously on a day when I was not available. I finally met Amory after the DPR was completed.

As I look back on a long career in Washington, DC, I can truthfully say that those six months were the busiest of my professional life.

We finally produced a draft that went to the cabinet-level Solar Energy Policy Committee for review and signoff. There was one bump in the road, some initial resistance by a DOE Deputy Secretary to the statement that the United States could achieve 20% renewable energy penetration by the year 2000 if the political will existed. After some discussion he did sign off and I personally delivered the report to Stu Eizenstat's staff at the White House on December 6. It was published formally as a U.S. government report in February 1979.

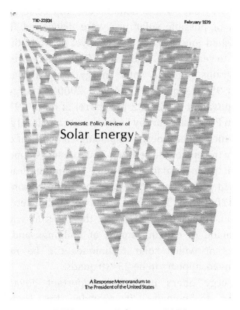

DPR report, February 1979

I reproduce below the major findings in the seven-page Executive Summary of the DPR for two reasons: It serves as a piece of history that most people today are not familiar with, and it shows the forward thinking that the DPR process produced 38 years ago. Many of its conclusions anticipated what is happening today as we begin finally to move toward a renewable energy future.

Domestic Policy Review of Solar Energy: A Response Memorandum to The President of the United States (February 1979, TID-22834/ Dist. Category UC-13)

EXECUTIVE SUMMARY

SUMMARY OF MAJOR FINDINGS

The results of the Domestic Policy Review can be summarized in nine major findings.

1. With appropriate private and government support, solar energy could make a significant contribution to U.S. energy supply by the end of this century. Renewable energy sources, principally biomass and hydropower, now contribute about 4.8 quads or six percent to the U.S. energy supply. Since estimates of future energy supply and demand are imprecise, three generic forecasts of possible solar use were developed. They can be distinguished most readily by the level of effort that would be required to reach them. In the Base Case, where present policies and programs continue, solar energy could displace 10–12 of a total of 95–114 quads in the year 2000 if energy prices rise to the equivalent of $25–32 per barrel of oil in 1977 dollars. A Maximum Practical effort by Federal, state and local governments could result in solar energy displacing 18 quads of conventional energy by the end of the century. Thus, if one assumes the higher future oil price scenario and this Maximum Practical effort, solar could provide about 20 percent of the nation's energy by the year 2000. The Technical Limit of solar penetration by the year 2000, imposed primarily by the rates at which changes can be made to existing stocks of buildings and equipment, and rates at which solar techniques can be manufactured and deployed, appears to be 25–30 quads.

2. Solar energy offers numerous important advantages over competing technologies. It provides the Nation with a renewable energy source which can have far fewer detrimental environmental effects than conventional sources. To the extent that increased use of solar energy can eventually reduce U.S. dependence on expensive oil imports, it can also improve our balance of payments, alleviate associated economic problems, and contribute to national security. Widespread use of solar energy can also add diversity and flexibility to the nation's energy supply, providing insurance against the effects of substantial energy price increases or breakdowns in other major energy systems. If oil supplies are sharply curtailed or environmental problems associated with fossil and nuclear fuels cannot be surmounted, solar systems could help reduce the possibility of major economic disruption.

In addition, because solar systems can be matched to many end-uses more effectively than centralized systems, their use can help reduce a large amount of energy waste. Although the U.S. now consumes about 76 quads of energy a year, less than 43 quads actually are used to provide energy directly in useable form. The rest in consumed in conversion, transmission and end-use losses.

3. Even with today's subsidized energy prices, many solar technologies are already economic and can be used in a wide range of applications. Direct burning of wood has been economic in the private sector for some time, accounting for 1.3 to 1.8 quads of energy use. Combustion of solid wastes or fuels derived from solid wastes is planned for several U.S. cities. Passive solar design can significantly reduce energy use in many structures with little or no increase in building cost. Low head hydroelectric generation is currently economic at favorable sites. Solar hot water systems can compete successfully in many regions against electric resistance heating, and will compete against systems using natural gas in the future. A number of solar systems installed by individual users are cost-effective at today's market prices. In addition, other solar technologies will become economic with further research, demonstration, and market development, and if subsidies to competing fuels are reduced or removed.

4. Limited public awareness of and confidence in solar technologies is a major barrier to accelerated solar energy use. Public testimony continually emphasized the need for more and better solar information. New programs to educate designers, builders, and potential solar users in the residential, commercial and industrial sectors are needed. Because consumers lack information, they often do not have confidence in solar products. Programs to provide reliable information to consumers, to protect them from defects in the manufacture and installation of solar equipment, and to assure competition in the solar industry can help build consumer confidence in the future.

5. Widespread use of solar energy is also hindered by Federal and state policies and market imperfections that effectively subsidize competing energy sources. These policies include Federal price controls on oil and gas, a wide variety of direct and indirect subsidies, and utility rate structures that are based on average, rather than marginal costs. Also, the market system fails to reflect the full social benefits and costs of competing energy sources,

such as the costs of air and water pollution. If solar energy were given economic parity with conventional fuels through the removal of these subsidies, its market position would be enhanced.

6. Financial barriers faced by users and small producers are among the most serious obstacles to increased solar energy use. Most solar technologies cannot compete effectively with conventional fuels at current market prices, in part because of subsidies, price controls, and average-cost utility rate structures for these conventional fuels. The tax credit provisions in the National Energy Act (NEA) will improve the economics of certain solar technologies, particularly in the residential sector.

   Other barriers exist because the high initial costs of solar systems often cannot be spread over their useful lives. Industry and consumers have yet to develop experience in financing and marketing solar systems. Some of the provisions of the National Energy Act will help expand credit for residential/commercial solar systems. In addition, the new Small Business Energy Loan Act will provide credit assistance to small solar industry firms. Other existing Federal financial programs, which were created for other purposes, could also help finance solar purchases if they were directed toward this end.

7. Although the current Federal solar research, development and demonstration (RD&D) program is substantial, government funding priorities should be linked more closely with national energy goals. Solar RD&D budgets, which have totaled about $1.5 billion in the Fiscal Year (FY) 1974 to FY 1979 period, have not adequately concentrated on systems that have near-term applications and can help displace oil and gas. Electricity from large, centralized technologies has been over-emphasized while near-term technologies for the direct production of heat and fuels, community-scale applications and low-cost systems have not received adequate support. Basic research on advanced solar concepts has also been under-emphasized, limiting the long-term contribution of solar energy to the nation's energy supply.

8. Solar energy presents the U.S. with an important opportunity to advance its foreign policy and international trade objectives. The United States can demonstrate international leadership by cooperating with other countries in the development of solar technologies, and by assisting developing nations with solar applications. Use of decentralized solar energy can be an

important component of development planning in less developed counties which do not have extensive power grids, and cannot afford expensive energy supply systems. In many cases, solar may be the only energy source practically available to improve rural living conditions. Through such efforts, the U.S. could also help to develop new foreign markets for U.S. products and services, thereby increasing opportunities for employment in solar and related industries at home. And, as solar energy eventually begins to displace imported oil and natural gas, the U.S. will enjoy greater flexibility in the conduct of its foreign policy. Insofar as solar energy systems reduce the need for nuclear and petroleum fuels in the long-term, they can help reduce the risk of nuclear proliferation and international tensions arising from competition for increasingly scarce fossil fuels.

9. Although the Federal government can provide a leadership role, Federal actions alone cannot ensure wide-spread use. Many barriers to the use of solar energy, and opportunities to accelerate its use, occur at state and local levels. In order to overcome these barriers and take advantage of these opportunities, a concentrated effort at all levels of government and by large segments of the public will be required. Nevertheless, the Federal government can set a pattern of leadership and create a climate conducive to private development and use of solar energy in a competitive market. These efforts must also recognize the wide variation among solar technologies and the resulting need to tailor initiatives to specific solar applications.

With the completion of the DPR my time at DOE was coming to a close, but this only became obvious a few months later. Chapter 5 discusses this post-DPR period.

Chapter 5

Post-Domestic Policy Review Period and the Ronald Reagan–George H. W. Bush Years

As mentioned in Chapter 4, the DPR was delivered to the Domestic Policy Staff at the White House on December 6, 1978, a date that my staff and I were more than pleased to reach. The work of the past six months had been demanding and challenging and helped to forge a strong bond among us. In fact, for several years afterwards when the group had effectively disbanded, those of us still in the DC area got together on that date to commemorate the experience and share memories.

In February 1979 the full report, Executive Summary plus Appendices, was published as a government report (TID-22834/ Dist. Category UC-13) which is available in the DOE Archives. That spring, President Carter made an important gesture of support for renewable energy by adding a solar hot water heating system to the White House roof. He dedicated it in a White House Garden ceremony in April, using the DPR as the foundation of his dedication speech. We renewable energy advocates were very pleased. The President followed up with a report to Congress on June 29, 1979, that described in detail his proposed renewable energy program. It outlined "...the major elements of a national solar strategy" and was based on the DPR. It showed that President

The U.S. Government & Renewable Energy: A Winding Road
Allan R. Hoffman
Copyright © 2016 Allan R. Hoffman
ISBN 978-981-4745-84-0 (Paperback), 978-981-4745-85-7 (eBook)
www.panstanford.com

Carter understood the importance of committing "...to a society based largely on renewable sources of energy" way back when. He deserves great credit for this foresight.

Then things started to come apart. Part of my group's responsibility was to help prepare the budget request for FY 1981, which was to be submitted to Congress in fall 1979. Based on the DPR findings we recommended an increase in the renewable energy R&D budget, which was denied by the Office of Budget and Management (OMB). We were told that the basis for the denial was the President's desire to balance the '81 budget and everyone had to share in the pain. As a loyal team member, I accepted this disappointment at first, but my attitude changed during the summer when, in an attempt to improve his deteriorating political standing, the President proposed an $88 billion program to develop synthetic transportation fuels (the so-called Synfuels Program).

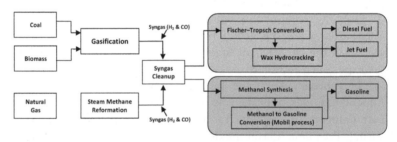

Indirect conversion synthetic fuels manufacturing processes

USD 88 billion was and still is a lot of money, especially in 1979, and I felt angry and betrayed. A few days later I left with my family for a vacation on Cape Cod, vowing that if I felt the same way when I returned in a few weeks I would resign my position. That happened a few months later in November.

My next stop was with the Energy Productivity Center of the Mellon Institute, a DC suburb-based part of Carnegie Mellon University. The center, headed by Roger Sant, a former senior official at the FEA, was focused on developing a Least Cost Energy Strategy for three major parts of the U.S. economy, industry, transportation, and buildings. I served as the Assistant Director for Industrial Programs. The center's work focused on the concept that energy is not important in itself, but only in terms of the end-use benefits or services (e.g., heating, cooling, lighting, communication, manufacturing, transportation) that energy makes possible. A driving belief for the Center's work was that the concept of energy services provides an important and useful framework for the development of national energy policy. More specifically, an energy policy framework that focuses on end-use services rather than fuels defines the technology and resource base for providing energy services most broadly. Generally, there is more than one way to provide an energy service and it would be in our individual and national economic interests to identify the least costly ways of providing these services. By doing so we free up financial resources for investment and economic growth.

An initial draft version of "The Least Cost Energy Strategy: Minimizing Consumer Costs Through Competition" was published in December 1979. It compared the actual U.S. energy experience in 1978 with a hypothetical 1978 experience that would have resulted from an application of least-cost principles, providing energy services at minimum life-cycle cost, during the previous 10–15-year period. The study suggested that had such a strategy been in effect prior to 1978, the United States in 1978 could have enjoyed the same level of energy services at reduced primary fuel consumption and lower cost. It further indicated that a least-cost approach can simultaneously address a number of objectives which, in the past, had been viewed as conflicting; continued economic growth, reduced imports, a cleaner environment, and lower consumer costs. A final, updated version of the Strategy was published in 1981.

In late February 1981, 14 months after I had left DOE and was working at Mellon, I was invited to testify before the Subcommittee on Energy Conservation and Power of the House Commerce Committee "...to discuss the development of our nation's solar energy policy, and to present my thoughts on how that policy should evolve in the future." My full testimony was placed in the record and included a review of the DPR process as well as my thoughts on priorities for national energy policy. My testimony made clear that the implementation of the DPR's recommendations was already in trouble for several reasons in addition to lack of adequate budget resources: lack of clear instructions to the federal bureaucracy on how to respond programmatically to the President's June 1979 message to Congress, failure to immediately establish a standing Solar

Subcommittee of the Energy Coordinating Committee to manage and coordinate the federal solar program, delay in assigning responsibility for the solar program within the DOE and appointing a Deputy Assistant Secretary for solar, and serious personnel shortages within the DOE for managing the solar program, resulting in "...serious morale problems, loss of program personnel, and strong and growing skepticism on the part of the public, the private sector, and other levels of government that the Federal government can be effective in speeding solar development."

A few months before my testimony, President Carter had been defeated for re-election by Ronald Reagan in the 1980 presidential election and the following eight years under President Reagan were difficult for renewable energy, and for the DOE in general.

President Ronald Reagan

The primary focus in the Reagan Administration was on nuclear energy and fossil fuels, as it would be again in the Bush-43–Cheney years, 2001–2008. In addition, President Reagan and his aides set out to eliminate two federal departments—DOE(1) (Energy) and DOE(2) (Education), but succeeded in neither due to resistance from a Democratically controlled Congress. Nevertheless, they did remove the solar panels from the White House roof and serious damage was done in those years to the renewable energy R&D budget. It was reduced by a factor of eight from its height at the end of the Carter Administration, from about $1B to $200M per year. Only the determined efforts of a few dedicated DOE managers, particularly Dr. Robert San Martin, the head of the renewable electric programs, kept the programs alive. It was also during this period that oil prices took a dive to below $10 a barrel and public interest in energy issues decreased significantly.

Things improved in the four years under Bush-41 (President George H. W. Bush), 1989–1992. R&D budgets edged up slightly and SERI was converted into and designated the National Renewable Energy Laboratory (NREL). Three other smaller renewable energy laboratories were also established, with suitable geographic distribution, as consolation prizes for other parts of the country that had competed to be the site of the first U.S. renewable energy national laboratory. The 1992 election also saw Bill Clinton elected as President and Al Gore as Vice President, and hopes were high that renewable energy would receive increased attention in the new Administration. I was now back at the DOE, having returned in 1991 to help run the renewable energy programs under San Martin, first as Associate Deputy Assistant Secretary, and then as Acting Deputy Assistant Secretary for more than three years. These and succeeding years are discussed in the following chapters.

To complete this partial history of the 1980s, with the completion of the work at Mellon in 1982, I joined the Congressional Office of Technology Assessment (OTA) for several months as a consultant to their energy program. During this time I was invited to join the National Academy of Sciences/National Research Council (NAS/NRC) staff as Executive Director of their Committee on Science, Engineering, and Public Policy (COSEPUP). Dr. Frank Press was then President of the NAS and he remembered favorably my work on his confirmation as Science Advisor to President Carter.

I accepted the offer from Dr. Press and was privileged to serve under COSEPUP chairman Dr. George Low, former senior official at NASA (he took over the moon-landing program after the fire that killed three astronauts on the ground in 1967) who was then President of Rensselaer Polytechnic Institute (RPI). He was a wise and gracious man.

George Low

My work at the NAS/NRC was to direct high-visibility public policy studies. These were focused on national security issues, education and employment issues, and an extensive series of briefings for President Reagan's Science Advisor, George Keyworth, on high-priority science and technology R&D needs. My responsibilities also included my organization of the NAS' celebration of the 200th anniversary of the French Revolution, which lasted a full week. Unfortunately, my COSEPUP portfolio did not include studies on energy issues, which were not a high priority for the Reagan Administration—NAS/NRC studies were almost always in response to government requests. This was also a period when energy costs were low—oil prices dropped below $10/barrel for a while in 1986—and public attention to energy issues was limited. Nevertheless, I followed energy developments closely, although at a distance from direct involvement. I eventually stayed at the NAS/NRC for more than eight years before returning to the DOE in 1991 for "a second tour of duty."

Chapter 6

The Clinton–Gore Years (Part 1 of 3)

In March 1991 I returned to the DOE as Associate Deputy Assistant Secretary for the Office of Utility Technologies (OUT), under the leadership of Dr. Robert (Bob) San Martin, which had responsibility for the Department's renewable energy electricity programs. I served as Bob's deputy. Responsibility for alternative renewable fuels such as ethanol and biodiesel rested in the Office of Transportation Programs, another part of our parent organization, the Office of Energy Efficiency and Renewable Energy (EERE). President George H. W. Bush was in the final two years of his presidency, and the 1992 presidential election was on the horizon. While SERI had become NREL and annual renewable energy R&D budgets did increase somewhat to about $300 million per year under President Bush, it was clear to me and others that more was required for a fully effective renewable electric program. This budget had to cover all aspects of photovoltaics (PV), concentrating solar power (CSP), wind power, hydroelectric power, geothermal power and ground-source (aka geothermal) heat pumps, biomass energy (wood, crops, animal wastes, algae), ocean energy (wave energy, ocean current energy, ocean thermal energy conversion (OTEC), energy storage (thermal storage, pumped storage, battery storage), and superconductivity (superconducting magnetic energy storage (SMES). I estimated the required annual amount to be $450 million.

The U.S. Government & Renewable Energy: A Winding Road
Allan R. Hoffman
Copyright © 2016 Allan R. Hoffman
ISBN 978-981-4745-84-0 (Paperback), 978-981-4745-85-7 (eBook)
www.panstanford.com

The 1992 election saw Bill Clinton elected as President and Al Gore as Vice President, and hopes were high that renewable energy R&D budgets would increase.

President Clinton

This was based on campaign statements that indicated the new Administration was more favorably disposed toward renewables than the two previous Republican Administrations. It was also based on a belief that both Clinton and Gore had a deep understanding of the nation's energy requirements and the need to make progress on developing and using energy efficiency and renewable energy. Both had had extensive interactions in their previous elected positions with the energy programs of the Tennessee Valley Authority (TVA), Clinton as Governor of Arkansas and Gore as a Congressman from Tennessee. However, I was not expecting much action in a first Clinton term as there were many issues on the new Administration's agenda and lots of other "fish to fry" after 12 years of Republican control of the White House. My hopes were more on actions related to energy in a second Clinton term. Of course my hopes were dashed when the President tried to put a price on carbon by raising gasoline prices by five cents a gallon and ran into a political firestorm. Unfortunately, he never tried again. Vice President Gore was also responsible for a serious setback when he insisted that all programs aimed at reducing global warming be so labeled in the FY1996 budget request, which many of us argued against strongly. Our fear was that with the Republicans winning both the House and Senate in the 1994 mid-term Congressional elections (the so-called Gingrich Revolution), such a guide would make it easy for Republicans to cut clean energy budgets. However, we were unsuccessful in the face of the Vice President's insistence and the Republicans subsequently used the "guide" to cut the requested OUT renewable energy budget by 25%. This had serious consequences for the NREL, which at that time received 60% of its operating funds from that budget, and the NREL was forced to lay off 200 of its 800 staff. It was a devastating time for renewables, about which I still carry strong feelings. One of those feelings is that we had a President and Vice President who understood energy issues and the need to move toward a renewable energy future. In my opinion they should have taken more steps 20 years ago to put us on that path and they didn't. I'm still upset.

In 1994 EERE's Assistant Secretary Christine Irvine left DOE and Bob San Martin became Acting Assistant Secretary under Secretary of Energy Hazel O'Leary. In an unusual move, Hazel (as everyone called her) had appointed a political appointee, Karl Rabago, as Deputy Assistant Secretary to be Bob's replacement. Historically, political appointments had been limited to the Assistant Secretary level and above. I then became Karl's deputy. Karl, a free spirit and someone comfortable in his own skin (an important characteristic of strong leaders in my opinion), had been a member of the Texas Public Utility Commission and a strong advocate of renewable energy. He and I worked closely together for a year until he decided that dealing with the bureaucracy was something he didn't enjoy. He left the DOE and I took over as Acting Deputy Assistant Secretary (DAS) for OUT, a position I held for the next three years. Karl and I are still close friends and he continues his work in support of clean energy.

Karl Rabago

In the following paragraphs in this chapter and in Chapters 7 and 8, I will discuss the status of the various OUT R&D programs during the 1990s when I had responsibility for these programs. They constituted the largest single part of the EERE budget and provided the basis for further technical developments and cost reductions in the years that followed.

Energy storage

My first act as DAS was to establish what I considered a real energy storage program, given my strong belief that energy storage was going to be critically important to the eventual widespread use of intermittent solar and wind energy. OUT's only energy storage program at that time consisted of an underground hot water storage project located at the University of Alabama, undoubtedly related to the fact that our budget requests had to pass through a House subcommittee chaired by an Alabama Congressman.

I also started an OUT battery storage program, tapping into expertise at several of DOE's national laboratories, initiated a superconducting magnetic energy storage (SMES) program, and looked to hydrogen as a long-term energy storage medium that used excess wind and solar energy to electrolyze water and produce hydrogen that could then be used as needed in fuel cells to generate electricity. The SMES program built on the discovery of so-called high-temperature ceramic superconductors in the mid-1980s that, when fabricated as coiled wires, could store energy in its magnetic field and discharge it quickly. While progress on commercializing high-temperature superconductors has been slowed by the difficulty of fabricating ceramic wires inexpensively, and the need to cool such superconductors to liquid nitrogen temperatures (77K or $-196°C$), many applications are under development, including SMES energy storage devices to compensate for voltage fluctuations in sensitive manufacturing processes, current fault limiters on power lines, zero-resistance, high-power cabling in electricity distribution systems, and high-power magnets in research and magnetic resonance imaging (MRI). To manage this expanded and increasingly technical program, I hired a PhD physicist. Today that program has been transferred to another part of the DOE and has taken on increased prominence with the recognition of energy storage's role in our emerging clean energy economy.

Database of State Incentives for Renewables and Efficiency (DSIRE): Another early decision was to set up an easily accessible online database that provided information on federal, state, and local government incentives available to consumers who wanted to install renewable energy systems. It had become clear to me that such a database was sorely needed. It was established in 1995 with funding from OUT and is still operating today under North Carolina State University. The database has since been extended to provide information on incentives for energy efficiency investments.

Distributed generation: Early in my tenure as DAS, my staff received a briefing from Karl Weinberg, Vice President for R&D at Pacific Gas and Electric (PG&E), a major utility in California, and two of his staff, Joe Ianucci and Dan Shugar. It described their new program to install large, local solar installations on their grid to reduce power line losses and forestall the need for expansion of existing substations or to build new ones. It was my first exposure to distributed generation in a utility setting and I was impressed. So impressed in fact that I immediately turned to Dr. Joe Galdo, a newly hired member of my staff and said that we needed a distributed generation program in OUT. That program was started the next day and for several years involved a close working relationship with the Electric Power Research Institute (EPRI), which was also interested in pursuing the distributed generation concept. The EPRI eventually decided to go on its own when they realized how important the concept was going to be.

An interesting offshoot of our program was my realization that my International Energy Agency (IEA) Renewable Energy Working Party colleagues were largely unfamiliar with the distributed generation concept. I arranged a briefing at our next meeting in Paris and arranged for both Joe Galdo and Joe Ianucci to join me for the presentation. It is now a concept used widely around the world.

Solar Energy

The solar energy program was well established when I took over OUT, well led by Bud Annon, and had the largest budget of any of the renewable electric programs. Efforts were focused on increasing sunlight-to-electricity conversion factors, reducing the manufacturing costs of solar PV cells and modules, increasing the reliability of DC-to-AC inverters, developing modules that could serve as building facades (Building-Integrated Photovoltaics, BIPV), and facilitating the financing of PV installations. Today, as PV module costs have come down dramatically, and progress is being made in reducing balance-of-system costs (support structures, wiring, permitting, labor costs), developing and providing attractive financing schemes has become the needed holy grail of PV deployment.

Solar PV array

It was also clear that the PV industry needed a boost if investments were to be made in expanded manufacturing facilities that would lead to reduced PV module costs. A large PV manufacturing facility at that time could only produce 50–100 megawatts peak (MWp) per year and resultant energy costs from PV were high, on the order of $1 per kilowatt-hour (KWh). It was also understood that without a domestic market as a foundation it would be more difficult for the PV companies to establish international markets. To address this need I hosted a meeting of U.S. PV company CEOs to propose and discuss a U.S. million solar roof program. After a long and fruitful discussion, it was agreed by all that such a program would be a good idea, I offered up some program money, and the industry got on board. The program started off well, continued for a while but eventually failed to achieve its objectives because of its inability to receive continued support from my successor as head of OUT.

Another part of the solar energy program addressed concentrated solar power (CSP). This technology used focused sunlight to heat water, create steam, and generate electricity via standard steam turbine generators. In the mid-to-late 1980s, 354 megawatts electrical (MWe) of parabolic trough solar concentrators were installed in Kramer Junction, CA, and are still operating today. They feed electricity into the southern California grid and benefitted in the early days from generous federal and state incentives.

Concentrating solar power—parabolic trough

Two other forms of CSP had also been under development for several years: power towers and dish-Stirling systems.

Concentrating solar power: power tower (left) and dish-Stirling system (right)

Power towers use a field of ground-mounted and adjustable sun-tracking mirrors. They surround a central tower that at its top has a receiver that is heated by the reflected sunlight from the mirrors. Dish-Stirling systems use individual curved radar-like dishes to focus sunlight on the heat-receiving end of a sealed Stirling engine to power an electricity generator. Both kinds of systems can reach the high temperatures needed for increased Carnot (heat to electricity) efficiency of the generators, and troughs and towers have the added and significant benefit of storing heat for power generation when the sun is not shining. Parabolic troughs use a black-colored oil to receive, transfer and store heat but the oil decomposes at temperatures just over 500°F. The power tower then under development at a test site in Daggett, CA (Solar Two), used molten salt as the heat transfer fluid and achieved temperatures over 1,000°F. Both the hot oil and the molten salt served as thermal storage mediums, allowing steam and power generation even after the sun went down.

With successful demonstration of Solar Two at a power level of 10 MWe, which was totally funded by OUT, I approached the industry about next steps. I proposed a cost-sharing program (50-50) as the logical next step toward commercialization, consistent with my belief that industry had to have "skin in the game" to expedite transfer of the technology to the marketplace. Industry balked, expecting continued full DOE funding, and after some difficult discussions in which no headway was made, I decided to terminate the program. Today power towers are receiving renewed attention both in the United States and in other countries, including several projects located in the California desert. The Solar Two site did not go to waste—after several years it was recommissioned as a site for astronomical observations using the multi-mirror field.

Wind energy

Wind farm: horizontal-axis turbines (left) and vertical-axis wind turbine (right)

The wind energy program was the second largest renewable electric program budget-wise, individual turbine sizes were then no larger than a few hundred kilowatts, and wind energy costs were about 45 cents per KWh. The program's focus was on blade design to increase energy capture, turbine design to increase conversion efficiency and equipment lifetimes, weight reduction to decrease investment costs, and studies related to bird deaths associated with turbine operations. The only U.S. wind farms were in Tehachapi, CA, again due to generous California state incentives. Most were horizontal-axis machines (rotors parallel to the ground) except for a few experimental vertical-axis machines (rotors perpendicular to the ground). Unfortunately, the first turbines in Tehachapi were installed along bird migration routes and large numbers of raptors and a few golden eagles were killed (or "taken" as we now say). It was soon realized that birds were not being sushi-ed by the turbine blades as they approached the blades but only after roosting on the horizontal struts of the turbine support structure and launching to go after identified prey. This understanding led quickly to the switch from strutted support structures to cylindrical support towers on which the birds could not roost and that are now in common use. Bird mortality associated with operation of wind turbines was a topic of some Congressional concern then and still is, along with concern about bat mortality more recently. In fact, at one budget hearing at which Assistant Secretary Irvine was testifying, the issue came up and she immediately turned to me and told the Congressman asking the question that Allan Hoffman will answer the question. As I walked up to the witness table, the Congressman labeled me the "bird man of DOE."

Wind energy technology was already under strong development in Denmark and was also just beginning to be recognized as an emerging technology suitable for utility application in the United States and in other countries such as China. In fact, during my visit to China in 1998 (see Chapter 10), I was taken to see China's only wind farm at that time, at a location north of Dalian and just below the Chinese-North Korean border. It had a few turbines, half of which were down for repair, but the Chinese program had the active support of its government and today China is the largest producer of wind turbines in the world. In this same vein, I should mention that in 1998 China was just beginning its PV program, again with strong government support, and today China is the world's largest producer of PV modules. It demonstrates the importance of government policy in advancing the development of renewable energy.

As with solar, our nation's wind development program was carried out largely at the NREL with OUT support. One issue that came to my attention during my 1999 visit to India, which was beginning to look seriously at wind and solar energy, was the fact that U.S.-manufactured turbines could not be sold in India at that time. Denmark, which had extensive experience with wind turbines and an aggressive export program, had convinced India to accept only European-certified turbines, a policy clearly supported by several turbine donations from the Danes to the Chinese. Upon my return from India, I asked the NREL to develop a competing U.S.-based certification system, this happened, and over time this problem went away. Sandia and Los Alamos National Laboratories also supported the wind program.

One other personal aspect of the U.S. wind program: In 1996, I visited the Tehachapi wind farm to see the installed turbines for myself as part of a trip to a renewable energy conference in California. One of the turbines I was shown was the 550 KWe Enron (previously Zond) turbine, just about the largest available at that time (a 750 KWe turbine was in development). I was accompanied by the head of the California Wind Energy Association (CWEA) and the actual designer of the turbine, Amit. It was about 140 feet tall from the ground to the bottom of the nacelle (about the height of a 10-story building), which supported the generator and the turbines blades. It was mounted on the old-style slatted support structure that has since been replaced by cylindrical support columns.

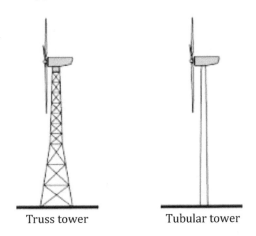

Truss tower   Tubular tower

Wind turbine support towers

For some crazy reason I decided I wanted to climb that tower even though I have a serious fear of heights—I guess I wanted to do it once in my life—and so informed my two companions. I will never forget the look on Amit's face when he asked me: "Are you serious?" He realized that he couldn't say no to the head of the program funding wind R&D back in DC and he couldn't let me climb alone. He would have to accompany me up and back down. Once it was clear I was serious, Amit and I were placed in safety suits that could attach to a cable running the full length of the tower and that could catch us if we started to fall. The only advice I received from Amit as we began the ascent was "Don't look down."

The climb up was not difficult and I did not look down. The scariest part was unhooking from the cable as we prepared to enter the nacelle, where we could stand up, and re-hooking to a support inside the nacelle.

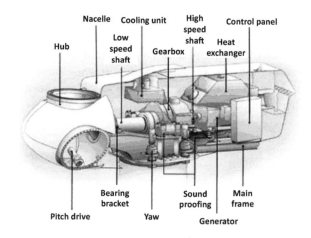

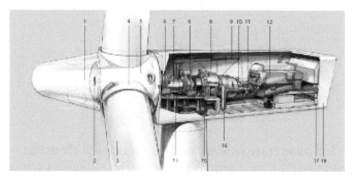

Wind turbine nacelle

The return trip went smoothly as well—I had had my once-in-a-lifetime experience. What I didn't know was that the CWEA Director on the ground was photographing the climb, and three weeks later I received unexpectedly a framed collage of those photos that graced my office wall until I retired.

Let me also mention that as wind energy began to attract more attention in the mid-1990s, the coal industry began to get nervous and produced a report that questioned whether wind could produce enough energy at a reasonable cost to provide a viable energy alternative. It was a clearly biased and misleading report that took a lot of my effort and that of my staff to refute. It also reflected the approach of the fossil fuel energy industry in its attempts to downplay the importance of emerging and potentially competitive renewable energy technologies—renewables are "cute" but just can't do the job. This is something I and others in the renewable energy field had to contend with for many years, and even to some extent today, a canard that was effectively disproven by the NREL's 2012 report entitled "Renewable Energy Futures Study" (NREL/TP-6A20-52409). This report concluded that "Renewable electricity generation from technologies that are commercially available today, in combination with a more flexible electric system, is more than adequate to supply 80% of total U.S. electricity generation in 2050 while meeting electricity demand on an hourly basis in every region of the country."

Finally, I should mention that at this time no attention was being paid to offshore wind, given the early stage of wind energy development, but which today I consider the most important emerging renewable electric technology. More on offshore wind in Chapter 12.

Chapter 7

The Clinton–Gore Years (Part 2 of 3)

This chapter continues the discussion of OUT technologies that began in Chapter 6.

Biomass energy

Biomass energy comes in many forms and is a large natural and renewable resource. Biomass resources include wood, which has long been used as a source of heat when burned, grasses and other crops, algae, municipal and industrial waste, and animal and human wastes (manure). Useful energy can be obtained from a number of these resources through direct combustion or conversion to alternative liquid and gaseous fuels. When I first took over responsibility for the renewable electric programs, this particular low-budget program was largely focused on wood. It supported efforts to develop fast-growing trees, more effectively combust wood and bagasse (sugarcane residue) for heat and biogas production, and mixing biomass fuel in the form of wood pellets or sugarcane residue with coal in power-generating boilers.

One significant change I introduced was the addition of a program to explore possible uses of animal wastes. The United States produces a great deal of such wastes, as do other countries, which present a serious disposal problem. The issue came to my attention when a pig farm with a large pit for liquid animal wastes overflowed its banks in Iowa and the resultant pollution of a nearby waterway made national headlines. A member of the

The U.S. Government & Renewable Energy: A Winding Road
Allan R. Hoffman
Copyright © 2016 Allan R. Hoffman
ISBN 978-981-4745-84-0 (Paperback), 978-981-4745-85-7 (eBook)
www.panstanford.com

EERE front office ran into me in a hallway and asked if we had a program to address animal waste issues. I answered honestly no, but immediately headed to my office, called the head of my biomass program, and directed that such a new activity be started. I also tasked one of the senior biomass program staffers to head up the new activity, which got underway the next day, and forever more he was known to me as "Dr. Poop" since he had a PhD.

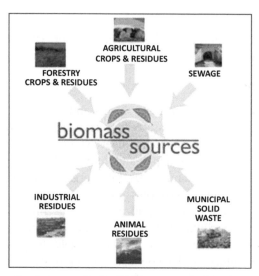

Admittedly, I knew little about animal wastes at that time but decided to get on a fast learning curve. To facilitate my learning, I contacted a colleague at Oak Ridge National Laboratory (ORNL) who worked closely with the University of Tennessee (UT) where animal waste issues were a subject of faculty research. I also learned that the U.S. Environmental Protection Agency (EPA) had a program on animal wastes because of the associated disposal and pollution problems. With his help, I scheduled an all-day meeting at UT where I could be briefed on the subject. To put it mildly I learned quite a bit about the amounts of animal wastes at issue (pigs and chickens are the biggest producers, although considerable contributions are made by turkeys and cows), the energy potential from digesting animal wastes and producing biogas, the ability to harvest unused drugs that are routinely fed to and pass through animals, and the uses of animal wastes for producing solid construction materials. I have never felt the same way about animal wastes again.

Of course animal wastes have been used as fuel for many years in the form of "cow chips" and as fertilizer derived from human wastes and known as "night soil." Small-scale biodigestion of wastes to produce cooking gas is routinely used in many countries, e.g., in many parts of rural China. China has also pioneered in large-scale biodigestion as I learned when I visited China in 1998. The U.S. delegation I headed for government-to-government meetings with the Chinese (there were seven of us) were taken to an experimental agricultural station just outside of Shanghai where we were shown three large stainless steel digestion tanks that were fed by the cow wastes of a nearby dairy farm. I always remember that the man who collected the wastes from about 3,000 cows with a shovel wore very high boots. He would bring the wastes to a device that separated the wastes from the straw in the cow bedding and then feed them into the tanks for several days of digestion in the presence of microbes. The resulting biogas, mostly methane, would rise to the top of the tanks and be filtered off to a pipeline that extended to the nearby town where the gas was used for cooking in both residences and restaurants. This reduced the need for burning charcoal with its many air pollutants. We were even taken to the town to see a home and restaurant where the gas was used.

You might ask what was done with the large amount of residue left in the tanks after the generated biogas was removed? Here the Chinese showed great ingenuity by selling the solid residue to the Netherlands for use as fertilizer in growing tulips. This still left a bit of liquid residue in the tanks which was sprayed on nearby fields to enhance crop growth. One of my thoughts on returning from that trip was admiration for the ability of the Chinese to use every bit of that waste.

One final thought on growing things in China: I happened to notice on the return trip from one planned visit to the countryside a number of plastic-covered shacks off to the side of the road. I asked my Chinese host to stop the several-car caravan to allow me to see what the shacks were and he resisted, saying it was nothing important. Nevertheless, I insisted and we stopped. Turns out that the plastic-covered shacks were high humidity hot houses where vegetables were grown for Beijing consumption, a high-income activity for the farmers. I subsequently found out that China uses more plastic sheeting than any other country. Such sheeting is much cheaper than glass for hothouse applications.

Hydropower

Hydropower, which taps into the kinetic energy of moving water, is renewable because of the recurring hydrological cycle where water vapor enters the atmosphere and returns as rainfall. In the mid-1990s large-scale hydropower was a well-established technology for producing electricity, using large dams in the western part of the United States equipped with multiple turbine generators. Such dams were also used for flood control, controlled irrigation, and recreation.

One problem facing such dams was the fact that too many of the fish present in the falling water passing through the turbines were being killed during the passage. Initial thoughts were that the fish were being cut up by the blades but research revealed that it was the pressure differentials created in the water by the blade movements that essentially caused the fish to explode. To overcome this problem, we started what I called the "fish-friendly turbine program" to design turbine blade arrangements that reduced these pressure differentials. Today such turbines are routinely available, along with fish ladders that allow fish to pass around rather than through the turbines.

Other issues faced by the hydropower program was the lack of an inventory of sites available for hydropower development, and resistance from environmental groups to further hydropower development for fish mortality and other ecological reasons. The United States has developed much of its large hydropower potential but more run-of-the-river facilities could be built. This latter resistance continues to this day. Other problems faced by hydropower sites are changing precipitation patterns that are associated with global warming and climate change. If dam reservoirs are not recharging due to limited rainfall, as is happening now in the U.S. southwest, power generation will be curtailed and less water will be available for agriculture.

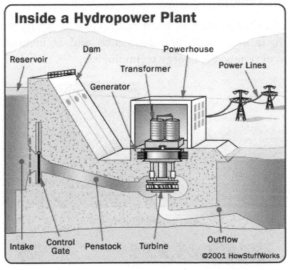

Inside a Hydropower Plant

©2001 HowStuffWorks

Another well-established aspect of hydropower is pumped storage in which water is pumped uphill at night using excess electricity to power the pumps and is allowed to run downhill during the day when peak power is required. In such facilities, which can only be built in suitable topographic locations, the turbine-generators act as both pumps and generators. This is an important but geographically limited form of energy storage.

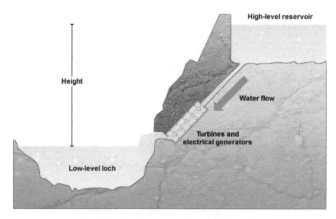

Sketch of pumped storage facility

Geothermal Energy

Geothermal energy derives from radioactive decay in the core of the earth, is essentially non-depletable and is included in the definition of renewable energy technologies.

Raw geothermal energy

Geothermal steam

Such energy manifests itself as steam or hot water arising from the earth and has been used throughout history in the form of hot water baths. In more recent times the hot steam has been used to power steam turbine generators—e.g., first in Larderello, Italy, in 1904, in the United States at The Geysers in California, and at other locations along the so-called volcanic ring-of-fire in other countries. Sufficiently hot water can be flashed into steam as well, or used to heat a volatile fluid such as Freon, which then powers a turbine-generator in a so-called binary power plant. Lower temperature hot water can be used to heat hot houses for agricultural purposes, and homes as is common in Iceland.

The DOE program I inherited was focused on reducing drilling costs for geothermal wells and developing a technology for extracting thermal energy from the hot dry rock resources that underlay all parts of the earth. These regions most often lie deep underground and have no associated water that can be heated into steam or hot water that can reach the earth's surface, but do represent an enormous energy resource. An effort to develop the technology to tap this resource had been underway at Los Alamos National Laboratory for several years and received its funding from OUT. As I became familiar with the program, I quickly came to appreciate its potential as well as the technological problems it faced. What was needed to tap into the thermal energy in the hot rocks was at least two holes drilled into the rock, with one hole bringing water down to the rock, the water then migrating through a fractured region connecting the two holes and absorbing heat, and the second hole directing the heated water to the surface. This was not an easy task as the holes have to be quite deep, techniques for fracturing of the rock at depth were still under development, water loss through the hole walls was a problem, and the environment in which the drilling had to take place was hot and often chemically reactive.

Nevertheless, the Los Alamos test facility demonstrated the technical feasibility of the concept and it was clear to me that the long-term future of geothermal was in the exploitation of hot dry rock resources. To confirm my belief I called a meeting in San Francisco with the CEOs of U.S. geothermal power companies and they quickly agreed as to where geothermal energy's long-term future lay. I then proposed a cost-shared R&D program and indicated I would issue an RFP (request for proposals) in which the DOE program would provide 50% of the needed funds. Unfortunately, the industry was in a critical cash-tight situation and unable to take on such a program when its basic survival was at stake and the RFP received zero proposals. Unwilling to support the program entirely with federal funds, with funding required to support other renewable electric technologies in a limited OUT budget, and the long-term nature of the energy payoff, I made the hard decision to terminate the Los Alamos program. I'm sure they still haven't forgiven me.

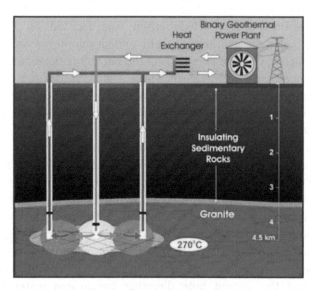

Sketch of hot dry rock facility

Another and smaller part of the geothermal program addressed geothermal (ground source) heat pumps. In my opinion this program belonged in the EERE Buildings Program, but my geothermal staff wanted to hold on to it and I saw no point in making a fight of it. After reviewing the program documents in detail I realized that the geothermal heat pump program was not in need of further R&D but rather a marketing program. It had the potential to reduce electricity demand for both heating and cooling for consumers and peak demand for utilities and was ready for full-scale commercialization. After talking with my Geothermal Program manager and identifying a staff member who could be assigned to the program, I approached the EPRI, the R&D arm of the private electric utility industry, to fund jointly over five years a $100M program to set up a new private sector organization, the Geothermal Heat Pump Consortium. This new organization would publicize the advantages of geothermal heat pumps and provide information and other assistance to utilities and consumers. The industry, via the EPRI, agreed to put up its $50M share and after some growing pains over the intervening years geothermal heat pumps today are an important energy efficiency option for both new and existing buildings. A major user of this technology is the U.S. Defense Department, which has to provide housing for its military service people and their families. In the late 1990s the U.S. Army installed 4,003 ground source heat pumps at Fort Polk in Louisiana, splitting the resulting cost savings for 20 years with the Energy Services Company (ESCO), which financed the original installation. Such third-party financing is now a routine part of energy efficiency projects, both in the military and in other sectors of our economy—e.g., public schools.

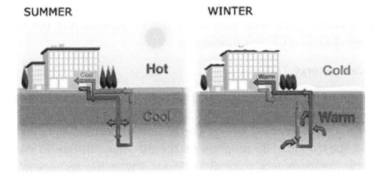

Geothermal heat pump systems for cooling and heating

An interesting offshoot of the U.S. program occurred during my first visit to China in 1998. I was there to oversee several programs that we had undertaken to assist China in its initial efforts at developing renewable energy technologies—solar, wind, and biomass. We had no agreements on geothermal, but as I traveled around China for 10 days, I was struck by the amount of new construction—building cranes everywhere. In fact, while preparing for my trip I was told that because of China's strong emphasis on infrastructure development the Chinese national bird should be the Building Crane. My informal reaction at every stop was to encourage the installation of geothermal heat pumps in their new buildings to get the full energy efficiency benefits over the life of the buildings. One of my ad hoc appeals was at Tsinghua University in Beijing (often described as China's MIT) and was noted by a Professor Li who contacted me as soon as I returned to the United States. He indicated his interest in starting a demonstration program in China and I agreed to support the effort as long as the heat pump equipment was purchased from an American company. Subsequently three different demonstrations were set up in different climate regions of China, including one in Beijing as part of the construction of a new high-rise apartment building. Instead of using heat exchange with the ground, the usual approach, this demonstration exchanged heat with an underground water source. On my second trip to China in 2001, I was invited to the dedication of this building and met the owner who was excited about his geothermal heat pump energy system. He said something to me that made quite an impression: "Every new building I build in China will include a geothermal heat pump." The dedication made the TV stations and the front page of a major Chinese newspaper.

Ocean energy

Ocean energy, which is getting lots of attention today, did not get much attention during the Clinton–Gore years except in one area—Ocean Thermal Energy Conversion (OTEC). Bob San Martin had started an OTEC program on the Big Island in Hawaii where ocean temperatures were suitable for trying the technology. In concept the technology is simple—in areas where the ocean surface temperature is high enough (85–90°F), one can run a low-efficiency (few percent) Carnot engine and generate electricity. The engine exhaust would be cooled by water pumped up from a reasonable depth, typically just above freezing. Given the large amount of thermal energy stored in the ocean even a small percentage of a large number can lead to significant power production. OTEC was also recognized as one of the continuously available renewable energy resources that could provide baseload power.

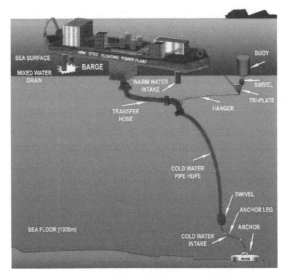

Sketch of mobile OTEC facility

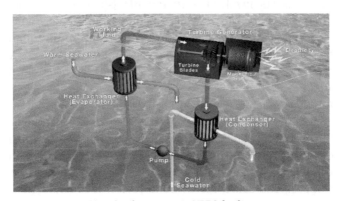

Sketch of stationary OTEC facility

Bob's program proved the technical feasibility of the concept via a 10 MWe test facility, but also pointed out the difficulties that would be encountered in moving to a practical power generating system. Heat exchange is an essential part of an OTEC facility and a large and efficient heat exchanger had to be maintained in an ocean environment with its many complications such as storms and surface deterioration due to sea life. The large structure required for heat exchange implied high costs, a long tube was required to tap into the deep cool ocean water (the first such tube deployed at the Hawaii facility came loose, sank, and was never recovered), and keeping the exposed surfaces clean required additional costs as well as attention to the polluting nature of the descaling fluids employed. The technology also was limited to certain ocean areas where the surface temperatures made economic sense. On the more positive side, aside from the baseload nature of OTEC, was the value of nutrients brought up with the cooling water, enabling aquiculture on land, and the fact that an OTEC facility could be mobile and moved to different locations during the year. Electrical energy could be transmitted to shore by cable if the facility was not too far offshore, but could also be stored as chemical energy in the form of ammonia or another compound produced using the generated electricity.

Funding for OTEC came up as an issue in the mid-1990s when I recommended canceling the Hawaii program because of the technology's elevated costs and limited geographical application. This did not sit well with the OTEC design community or the scientists involved with the Big Island test facility, and especially not with the Hawaiian Congressional delegation. This was especially true of Senator Inouye. He kept putting funds for the facility back into the budget even though I had zeroed out the request, and this went on for a few budget cycles. He finally relented.

Another Senator, who chaired the Senate Energy Committee, Sen. Bennett Johnston, Jr. of Louisiana, had a constituent who wanted to build mobile, ship-based OTEC facilities. Not being in a position to turn down a request for further evaluation from a powerful Senator, I met with the constituent and his staff, received his proposal, and asked the NREL to review it in detail. After this review our conclusion was the same as before, too expensive and problematic to justify DOE support. Sen. Johnston was kind enough to drop the issue at that point.

One further point about OTEC: It comes in open-cycle and closed-cycle versions and the open cycle produces distilled water that for isolated island locations may be more valuable than the electricity produced. Several small OTEC facilities exist today, and the U.S. military has expressed interest in OTEC for some island bases, but no commercial facilities currently exist.

The term "ocean energy" also encompasses several other energy technologies—wave energy, tidal energy, and ocean current energy. The first converts the energy in rising and falling waves into mechanical motion that is then used to generate electricity.

Sketch of wave energy units

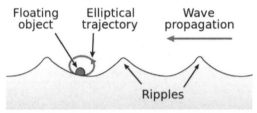

Wave energy dynamics

Little attention was paid to wave energy in the mid-1990s because of costs and the fact that the U.K. had, early on, put several large wave energy machines into the rough environment of the ocean and they had all been destroyed. It was not until researchers started developing and testing smaller buoy-type devices that wave energy began to get serious attention and today wave energy is an important emerging technology that offers near baseload capacity. Early deployments are taking place in Scotland and Australia, the U.S. Navy is supporting research efforts, and plans are being made for initial commercial deployments off the U.S. west coast.

One form of tidal energy takes advantage of the fact that at certain locations the difference in ocean height between high tide and low tide is quite large. This allows the water captured at high tide to be released to run downhill at low tide and thus create a hydropower facility. It was an established but geographically limited technology when I took over responsibility for the ocean energy program and it received no support from my limited budget. Another form of tidal energy taps into the kinetic energy in moving masses of water as tides go in and out daily on a regular basis—e.g., the tides in New York City's East River. I still remember the day near the end of my tour of duty as head of OUT when a representative of the Verdant Power Company came to my office and suggested that we put a "wind machine in the East River." Like an ordinary wind turbine, this machine would tap into the kinetic energy of a moving fluid, reversing direction as the tides reverse, but instead of air this turbine would tap into the kinetic energy in water. The water's relatively low speed would be compensated for by the mass of the fluid moving past the propeller blades. I was unable to do more then to pursue the technology, but today a test system in the East River is generating power and delivering it to New York City.

Verdant Power tidal energy turbine

A third form of ocean energy taps into the kinetic energy of constantly moving "rivers" in the ocean that exist below the ocean surface off of various coastlines—e.g., the Gulf Stream that flows along the U.S. east coast. This is another geographically limited power source but it represents a large potential energy resource and would provide something close to baseload power. Sometimes, over time, these underwater rivers change direction, requiring a movement of seafloor-anchored turbine arrays. This technology requires further development to reduce costs and increase reliability, and suffers from the difficulties all ocean energy technologies experience by being in the difficult ocean environment.

Hydrogen

The OUT program also had responsibility for developing cost-effective technologies for producing and storing hydrogen. As a physicist it has been an article of faith for me to believe that mankind will eventually use large quantities of hydrogen as an energy source because it is the most abundant element in the universe. On a practical level hydrogen is the fuel, along with oxygen, that is needed to power fuel cells which produce electricity, water, and heat.

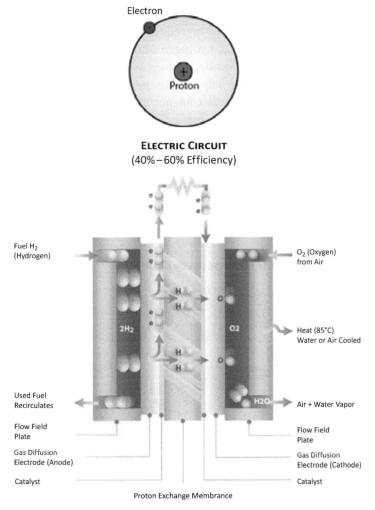

Hydrogen atom (top) and sketch of a fuel cell (bottom)

It can also be burned cleanly and can serve as an energy storage medium for excess electricity from solar, wind, and other generating technologies. This latter application would use excess electricity not needed at the time (e.g., at night when the wind is blowing most strongly) to power electrolyzers that break down water, H_2O, into hydrogen and oxygen. Unfortunately, electrolysis costs have historically been too high for commercial applications of hydrogen (e.g., petroleum cracking and food hydrogenation) and most hydrogen today is produced by steam-reforming of natural gas. This process puts carbon dioxide into the atmosphere unless it is captured and sequestered, a process still under development. Breakthroughs in reducing electrolyzer costs are needed to bring electrolysis hydrogen into widespread use, along with attention to perceived safety issues associated with hydrogen storage (the Hindenburg effect). My personal view is that if hydrogen is used in personal transportation vehicles that utilize hydrogen-burning engines or fuel cells that power electric drive motors, that would be safer than sitting on a tank of gasoline or diesel fuel as we do today in traditional internal combustion engine cars and trucks. Of course providing widely available hydrogen for these applications is not a trivial or inexpensive task and onboard hydrogen storage is likely to be overtaken by recent developments in battery storage for electric vehicles.

Model E Crossover: Fuel Cell

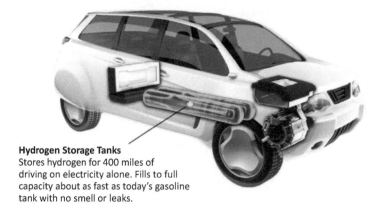

Hydrogen Storage Tanks
Stores hydrogen for 400 miles of driving on electricity alone. Fills to full capacity about as fast as today's gasoline tank with no smell or leaks.

Given hydrogen's significant but perhaps long-term potential as an energy source (technically, it is an "energy carrier" as it is a derivative fuel), I was a strong supporter of OUT's hydrogen R&D efforts. I managed to get the hydrogen budget up from $1.8M in 1993 to $20M a few years later. In addition to efforts on reducing the costs of electrolysis, considerable effort was put into hydrogen storage, particularly in high-pressure storage containers (5,000–10,000 psi) that would be needed in cars if sufficient hydrogen in individual fuelings were to be available for ranges of several hundred miles. Chemical storage of hydrogen in solid materials was also actively explored. These issues continue to be the focus of R&D efforts today.

Chapter 8

The Clinton–Gore Years (Part 3 of 3)

In 1997 Bill Richardson was appointed Secretary of Energy, he brought several new people into the Department, and I was replaced as head of OUT. My new assignment, along with a few other "displaced" senior managers, was to work on climate change issues, which were beginning to attract some attention. My background in renewables was clearly an asset in this work and my new group was active in developing and supporting clean energy initiatives with developing countries. It was satisfying work, and freed from the responsibilities of running a large organization with a large budget, I even had time to write a long chapter on solar energy for a new energy encyclopedia.

In mid-August 1999 I was approached by a representative of the DOE's Policy Office, Gene DeLatorre, who asked me to represent the United States at a meeting in Amman, Jordan, only a few weeks later in mid-September. The purpose of the meeting was to help plan a high-level water conference to be held that December in Jordan that would involve the United States as well as King Hussein of Jordan, Yasser Arafat, President of the Palestinian Authority (PA), and Ehud Barak, Prime Minister of Israel. You can imagine my surprise at the invitation as I had no professional involvement with water issues at that time, and my reaction to the invitation was "Why me? I don't know a damn thing about water except what I read in the papers." Of course, at that time President Clinton was engaged deeply in peace

The U.S. Government & Renewable Energy: A Winding Road
Allan R. Hoffman
Copyright © 2016 Allan R. Hoffman
ISBN 978-981-4745-84-0 (Paperback), 978-981-4745-85-7 (eBook)
www.panstanford.com

talks between the Israelis and Palestinians and it was well known that water issues were high on the agenda of the peace negotiations. Gene explained to me that I had three qualifications the Policy Office was looking for: I was a senior DOE official, I had a strong background in renewable energy which everyone recognized would play a role in future desalination efforts in the Middle East, and I had extensive experience in meeting with senior officials of other governments. Not really being in a position to say no I accepted the invitation and put myself on a steep learning curve with respect to water issues. This involved lots of reading, especially with regard to desalination and its energy requirements, and meetings with people in the DC area who had relevant Middle East experience. After several weeks of this preparation I left for Amman where I was scheduled to meet up with two water experts from Livermore National Laboratory as part of a small U.S. delegation at the meeting. That time in Amman changed my professional life for the next 12 years.

The meeting in Amman, organized by a former U.S. Congressman from Salt Lake City, was part of an effort to educate American legislators about Middle East issues and improve relations between Israel and the Palestinian Authority. About 50 people participated, mostly Israeli and Palestinian water experts, and a few Americans, including an expert on Middle East water issues from the United States Department of State. It was for me an enlightening experience—I had not previously appreciated fully the nature of water problems in the Middle East, the importance of energy in addressing water issues, and the critical role of water in the peace negotiations. In a defining moment for me, I was deeply impressed by a statement made by Nabil Al Sharif, head of the Palestinian delegation, the PA's chief water negotiator, and a civil engineering classmate of Yasser Arafat. He stated: "There will be no peace in the Middle East until the water issue is addressed."

(One interesting side note about Nabil: When I visited the chief Israeli negotiator on water issues in his office at Haifa University, I couldn't help noticing a photo of Nabil on his wall. Turns out they were good friends.)

Eventually the peace negotiations faltered because of other issues and the anticipated high-level water meeting of principals never took place. In any event I returned to the DOE with an increased interest in water and water-energy issues and vowed to learn as much as I could to explore what was for me a new area of professional interest. It is fair to say that this new interest helped define my final years as a federal energy official.

Nabil Al Sharif—Palestinian Authority Water Minister

As I ventured into this new subject, I quickly came to understand that water and energy issues are intimately linked—energy is needed to produce and deliver clean water and water is needed to produce fuels and energy. In an article I wrote a few years later, I labeled this relationship the water-energy nexus.

I also came to understand that many of the things I had been saying in my writing and public presentations about energy could also be said about water. Namely, there is no shortage of energy (water) in the world, but there is a shortage of inexpensive energy (clean water) that people can afford to buy. One significant difference is that there is no substitute for water, whereas energy comes in several different forms and substitute energy systems do exist.

After about a year I felt confident enough to give my first public talk on water-energy issues in 2001 at an Organization of American States (OAS) meeting in Washington, DC. In the years since I have spoken often and published widely on the subject. This is despite the fact that there was high level resistance within EERE to addressing the water-energy issue, partly because of the possibility of diverting funds from other EERE activities, sometimes referred to as "mission creep." This made things difficult and professionally a bit lonely, and I have always believed that it was only because of the fact that I was a senior DOE official who had more-than-average control of his schedule that I was able to pursue my water-energy interests.

In August 2004 I was invited to write a paper on the nexus, which I entitled "The Connection: Water and Energy Security" and which was published in the newsletter of the Institute for the Analysis of Global Security (IAGS). The article attracted considerable attention and several speaking invitations. It began by stating: "The energy security of the United States is linked to the state of its water resources. No longer can water resources be taken for granted if the U.S. is to achieve energy security in the years and decades ahead. At the same time, U.S. water security cannot be guaranteed without careful attention to related energy issues. The two issues are inextricably linked." To quote further from that article, I explained the linkage between energy security and water security as follows:

> Energy security rests on two principles—using less energy to provide needed services, and having access to technologies that provide a diverse supply of reliable, affordable and environmentally sound energy. Many forms of energy production depend on the availability of water—e.g., the production of electricity at hydropower sites in which the kinetic energy of falling water is converted to electricity. Thermal power plants, in which fossil, nuclear and biomass fuels are used to heat water to steam to drive turbine-generators, require large quantities of water to cool their exhaust streams. The same is true of geothermal power plants. Water also plays an important role in fossil fuel production via injection into conventional oil wells to increase production, and its use in production of oil from unconventional oil resources such as oil shale and tar sands. In the future, if we move aggressively towards a hydrogen economy, large quantities of water will be required to provide the needed hydrogen via electrolysis.

> Water security can be defined as the ability to access sufficient quantities of clean water to maintain adequate standards of food and goods production, sanitation and health. It is of growing importance because the world is already facing severe water shortages in many parts of the developing world, and the problem will only become more widespread in the years ahead, including in the U.S. Just as energy security became a national priority in the period following the Arab Oil Embargo of 1973–74, water security is destined to become a national and global priority in

the decades ahead. Central to addressing water security issues is having the energy to extract water from underground aquifers, transport water through canals and pipes, manage and treat water for reuse, and desalinate brackish and sea water to provide new water sources.

In the years since I have written much and spoken often on the subject, and the water-energy issue has slowly emerged from the woodwork and is today considered a hot topic. Even DOE headquarters has begun to pay attention and recently issued a report on the energy-water connection. It should also be acknowledged that several DOE national laboratories had been involved in studying aspects of the issue for several years. What was lacking was any headquarters funding and coordination of these efforts, which the laboratories individually pursued using discretionary dollars. This coordination finally began at a regular meeting of senior laboratory and DOE officials at which I introduced the water-energy topic and asked for each laboratory representative to describe any activities already underway. Five minutes per lab had been set aside on the early afternoon agenda for such comments, which proved to be inadequate. The rest of the afternoon was devoted to these comments, and at the end of the meeting the laboratory representatives met and agreed to a formal coordination of these efforts under the leadership of the Sandia and Berkeley laboratories. One outcome was a series of water-energy briefings for senior DOE officials by laboratory teams.

My published work came to the attention of Professor Gustaf Olsson of Lund University in Sweden, who during a trip to the United States in 2006 stopped in to see me and learn more about the water-energy nexus. He became quite interested, went on to become an expert in the field, and in 2012 published *Water and Energy: Threats and Opportunities*, an encyclopedic work on the subject. Its second edition appeared in 2015. Prof. Olsson and I have remained in close contact over the years and co-authored, with Andreas Lindstrom, a long report on the water issues associated with fracking ("Shale Gas and Hydraulic Fracturing—Framing the Water Issue," Swedish International Water Institute, August 2014).

Mention should also be made of several other U.S. government water-energy developments. In the mid-1990s a multi-agency agreement had been signed by the DOE, the DoD, the EPA, and the

Department of State to cooperate on international environmental issues. The DOE's and the EPA's interest in environmental issues may be understandable, but why the DoD? This reflected the DoD's wise recognition that issues involving energy and water could have implications for DoD involvement in international areas of tension—e.g., protecting our access to petroleum supplies from Persian Gulf countries, tensions among countries competing for fossil fuel supplies, and water shortages that create transboundary conflicts and migratory pressures that have the potential to threaten U.S. national security interests. As the DOE's representative for this agreement, I was impressed by the DoD's prescience about and commitment to these issues. It remains an important item on their agenda.

Another activity focused on the linkage between water and energy was the solar-powered desalination project we undertook in Jordan in cooperation with the U.S. Agency for International Development (AID) starting in 1997 ("Solar Powered Desalination and Pumping Unit for Brackish Water"). AID had a major program for developing clean water resources in developing countries. The DOE was also focused on water resources in the Middle East as described above, and we proposed a joint project with AID in which they would provide financial resources and the DOE would provide some financial support, technology guidance, and manpower for a solar-powered desalination project in Jordan. It was based on the fact that many arid and desert-like regions with large populations, such as the Middle East and North Africa (MENA) suffer from scarcity of fresh water, which is needed for drinking, sanitation, and agriculture. Often the only water resource is salt-contaminated underground brackish water with salinity levels in the range 3,000–8,000 parts per million (ppm) of total dissolved solids.

Several studies had identified a number of small communities in the Palestinian West Bank, and many more in Jordan, that lacked access to both electricity and potable water, and it was recognized that if energy were available to bring brackish water to the surface and reduce salinity levels to World Health Organization standards for potable water (less than 1,000 ppm), one could address these needs and help generate sustainable economic development. It was also recognized that the MENA region is rich not only in brackish water but also in solar energy. Thus, it was only natural to think about a project in this region to demonstrate, in several small villages, the feasibility of operating small, solar-powered units for desalinating brackish water.

The village chosen by the Jordanian Ministry of Water and Irrigation for the original desalination demonstration was Qatar, a small village of 250 residents 20 miles from the coastal city of Aqaba. The village had no fresh water supply (fresh water was trucked in from Aqaba on a weekly basis) but did have two wells drilled into a brackish water aquifer (depth: 50 meters; salinity: 3,865 ppm). The desalination unit was a reverse osmosis water purification unit (ROWPU) declared surplus by the U.S. Army. We had it refurbished and shipped to Jordan by sea. It was powered at first by a diesel generator to gain experience with the desalination process, with the goal of eventually replacing the diesel generator with a PV system. The Jordanian National Energy Research Center (NERC) agreed to install and maintain the system. Day-to-day U.S. project responsibilities were assigned to Dr. Ken Touryan at the NREL.

With the lessons learned from this first installation, which was eventually connected to the Jordanian power grid when it was extended from Aqaba, the NREL and the NERC collaborated on a design for a solar-powered desalination unit. It was installed at an industrial park on the outskirts of Aqaba at the request of Jordan's King Abdullah, who wanted to demonstrate the desalination unit to his people. The most challenging design problem was matching the intermittent PV power supply output with the steady energy demand of the desalination process. The final system used a desalination unit capable of producing 15 gallons per minute and powered by a 16.8 KWp PV system backed up by a battery storage system. A third demonstration was planned for a small village on the Palestinian West Bank but was never implemented because of political difficulties between the PA and Israel. Nevertheless, this set of experiences showed that solar-powered desalination of brackish water was a viable option for the MENA region and similar regions around the world, and that adversaries could work together to meet mutual needs. It is an important application of renewable energy (of course any source of electricity could have been used to power the reverse osmosis units) and will become even more important in the future as the demand for desalinated water grows in both developing and developed countries.

Village of Qatar

PV panels and brine evaporation pond

The Clinton–Gore years came to an end with Al Gore's loss to George W. Bush (Bush-43) in the 2000 presidential election. The new Administration's approach to energy issues is discussed in Chapter 10. However, before we leave our discussion of the 1990s, I want to mention an important event for renewable energy in 1996, the Summer Olympic Games in Atlanta, Georgia. This unique exposure of renewable energy to the Game's world TV audience is discussed in the next chapter, Chapter 9.

Chapter 9

The 1996 Summer Olympic Games in Atlanta

Installing PV panels with Atlanta in the background

One renewable energy event that could be lost in the stream of history is the DOE's role in the 1996 Summer Olympic Games held in Atlanta, Georgia. It was an important event in the sense

The U.S. Government & Renewable Energy: A Winding Road
Allan R. Hoffman
Copyright © 2016 Allan R. Hoffman
ISBN 978-981-4745-84-0 (Paperback), 978-981-4745-85-7 (eBook)
www.panstanford.com

that it set a new standard for the greening of the Olympics, a standard that has been met in subsequent Summer Olympic Games. This includes the 2000 Games in Sydney, the 2004 Games in Athens, the 2008 Games in Beijing, and the 2012 Games in London. It is expected that the 2016 Games in Rio de Janeiro will meet this standard as well.

When Atlanta was selected for the 1996 Summer Olympic Games it was understood that it was likely to be the last U.S. city to host the Summer Games until well into the 21st century. Planning for the DOE's activities at the 1996 Summer Games began in 1990, under the Bush-41 Administration, at the DOE's Atlanta Support Office. It took formal shape in March 1992 when the newly formed National Renewable Energy Laboratory and the Atlanta Support Office formed a team to identify and discuss opportunities with the Atlantic Committee for the Olympic Games (ACOG) and the Metropolitan Atlantic Olympic Games Authority (MAOGA). Technical opportunity teams were formed, and initial discussions with industry and other potential stakeholders began that fall. The DOE's interest was clear—the Olympics represented a unique and highly visible opportunity to showcase American progress in developing energy efficiency and renewable energy technologies and speak directly to potential global markets. More than two million on-site visitors were expected in Atlanta, as well as a global TV audience of more than three billion.

Unfortunately, much of this planning activity got lost in the aftermath of the 1992 U.S. presidential election, the beginning of the Clinton–Gore Administration, and the appointment of a new Secretary of Energy. This became clear at a meeting I attended in February 1993 that was called by the new Secretary, Hazel O'Leary, to discuss her plans for the future. There was no mention of the planning for Atlanta, which some of us were familiar with, but we just assumed that a new Secretary has lots of issues to think about.

When it became clear shortly thereafter that no active planning for Atlanta was still underway, and that no DOE funds had been budgeted for activities in Atlanta, two of us decided to act—John Millhone, head of the EERE Buildings Program, and me. Under my leadership we revived the effort as an ad hoc activity that would be carried out using program funds over which John and I had discretion, and, hopefully, generous in-kind and cash contributions from private sector partners who wanted to demonstrate their technologies. On this basis a significant demonstration project was planned and implemented. Final estimated cost of the project was $25 million (in 1996 dollars) with the DOE's share being $5 million. Getting the $5 million out of our limited budgets was not easy. I had to twist arms of two people—one in my own shop, and John's. John was fully supportive of the project but felt so badgered by my repeated requests for funds that after a while he would occasionally turn the other way when he saw me approaching him in the hallway. Nevertheless, he came through and today a building in Atlanta exists because of John's financial support. More on that later.

The issue in my shop was with the head of my solar energy program, Bud Annon's successor Jim Rannels, who claimed all his allocated funds were already committed to other projects. While I could have pulled the money in my role as DAS, I did not want to do that without first talking with him. The funds I wanted were to be used for a large solar PV installation on the roof of the new 1-million gallon Olympic swimming pool (Natatorium) that was being built on the Georgia Tech campus for the Games. After a bit of delicate and some not-so-delicate persuading, he saw the value to his program and agreed to adjust his budget.

To ensure the widest possible visibility of our clean energy projects in Atlanta, I decided to hire a contractor, Casals and Associates, to develop a detailed outreach plan that would take full advantage of the Olympic opportunity. We worked closely together with the contractor team for the next two years, producing a detailed plan that would give us good TV exposure and relied on partnerships with small and large corporations. A copy of the plan still resides in my files. Unfortunately, 1996 was the year of Secretary O'Leary's infamous trip to India which raised Congressional concerns, and EERE's leadership, being gun-shy about stirring up the already stirred up Congress, directed me not to implement several aspects of the outreach plan. As a result parts of a unique opportunity were lost. Nevertheless, several clean energy demonstrations were in place on the Georgia Tech campus during the Games and several provided ongoing benefits to Georgia Tech and Atlanta after the Olympic Games were concluded.

The full planning and implementation took three years. What was accomplished? First and foremost, many of the two million people attending the Olympics saw renewable energy technologies at work up close. In addition, millions more saw TV coverage of the technologies posted by the media—not as many as we had hoped because of the restricted outreach activity, but still a lot. What didn't take place, e.g., was an advertising campaign planned jointly with the Coca-Cola Company, which has its headquarters in Atlanta. We also lost out on a solar hot water heating demonstration for the equestrian stables. We learned that the horses are washed every day with 90 degree Fahrenheit water, an easily obtained temperature for solar hot water heaters. An agreement was almost in hand with the organizers of the Games when the local utility offered to heat the water electrically at a very low cost. Since this was a cost well below market costs, and solar hot water heating systems were still not well known, the organizers went for the electrical heating.

Nevertheless, we managed to introduce quite a few clean energy demonstrations into the Games, including

- 40,000 square feet of the roof area of the newly built Natatorium were covered with 2,856 photovoltaic (PV) modules, delivering 340 kilowatts of peak electrical power to the swimming complex. At the time, and for several years afterwards, this was the largest PV building installation in the world. It is still operating under the guidance of the Georgia Tech electrical engineering department and is used as a teaching tool as well as a source of electricity.
- 274 solar pool heating panels were also mounted on the Natatorium roof, to heat the one-million gallon Olympic pool.
- A 9 kilowatt peak PV array was mounted over the walkway to the Natatorium, which charged a battery storage unit to offset nighttime lighting electrical needs.
- Several hundred alternative fuel vehicles, both buses and light duty vehicles, were used as part of the Olympic fleet. Fuel sources used included compressed natural gas, liquid natural gas, battery-stored electricity, and hydrogen.
- A stand-alone PV-powered outdoor lighting system (65 double-lamped fixtures) provided the illumination for the

visitor parking lot of the National Park Service's newly built Martin Luther King, Jr. Visitor Center.

- A 7 kilowatt peak dish-Stirling engine, multi-faceted, concentrating solar power unit was on loan to Georgia Tech for demonstration during the Olympics.
- A local school roof was selected for installation of a highly reflecting "cool roof" to reduce cooling requirements.
- A building on the Georgia Tech campus and the newly built Southface Energy and Environmental Resource Center were selected as sites for geothermal heat pump demonstrations. The 6,000 square foot Center, located on land leased from the City of Atlanta and located next to Atlanta's Science Museum, was built with funds from EERE's Buildings Program as a demonstration site for current or emerging energy efficiency and renewable energy technologies for the building sector. Southface Energy Institute is an environmental consulting group that since 1978 has worked with the construction and development industry, government agencies and communities to promote sustainable homes, workplaces and communities.

Natatorium roof with installed PV and pool heating panels (left) and natatorium pool under the solar roof (right)

All of this could not have been done without private sector contributions. The PV panels were provided and installed at cost by the Solarex Corporation, the pool heaters were donated and installed by an industry trade association on a cold and wintry day, and Atlanta provided a free 50-year lease for the new Center which the DOE donated to Southface.

Two side notes to the Atlanta story. One was a visit from the organizers of the Sidney, Australia, 2000 Summer Games, who wanted to learn more about what we had done in Atlanta. We shared information on our approach to clean energy demonstrations, and they did us proud when it was their turn four years later. The City of Atlanta had also provided the natural gas refueling station for the Olympic buses and vans and used it to initiate a large-scale conversion of municipal and other vehicles to natural gas.

After the Games were completed, I was asked by the DOE Atlanta Office what I would like as a thank you for our efforts with Atlanta. I requested honorary Atlanta citizenship, which I received in a framed certificate signed by the Mayor of Atlanta. The head of the Atlanta Office, Jim Powell, a good old southern boy who knew I grew up in New York City, presented the certificate to me with the following words: "You may now be a citizen of Atlanta but you'll always be a Yankee." True enough.

Chapter 10

Cooperation with Other Countries to Develop Renewable Energy Technologies

It was clear to some of us in the 1990s that a transition was coming from reliance on fossil fuels to an energy system increasingly reliant on cleaner sources of energy. Nuclear power had its strong advocates, but a number of us preferred a different path, renewable energy. We believed that it was possible with adequate policy and financial support and were making every effort to speed up the steps leading to the transition. This was clearly an issue not only for the United States but for other countries as well if the world was to eventually reduce dependence on coal, oil, and natural gas. These fuels, when combusted, put carbon and other pollutants into the environment, created international tension via competition for fuels, presented variable and often volatile costs, and impacted national security. Methane (natural gas) was also identified as a powerful greenhouse gas in its own right.

This translated into a desire to work with other countries to advance their renewable energy programs and the EERE front office undertook an effort to do so. Unfortunately, this effort did not go well and after a particularly disheartening meeting on the subject in the Assistant Secretary's conference room, I returned to my office determined to set up an international program within OUT. The first step was picking up the phone,

The U.S. Government & Renewable Energy: A Winding Road
Allan R. Hoffman
Copyright © 2016 Allan R. Hoffman
ISBN 978-981-4745-84-0 (Paperback), 978-981-4745-85-7 (eBook)
www.panstanford.com

calling Dr. Samuel Baldwin at the NREL, who had some international experience, and with whom I had worked previously and respected highly. I asked him to start thinking about how such a program should be structured and decided to take 1% of my budget of about $300M to support such an effort. This was within my budgetary discretion.

The results of this decision were cooperative programs with China, India, and Israel, all of which were aggressively exploring possibilities for renewables, and a few smaller efforts with other countries. We also began discussions with the European Union (EU) Commission's Development Program for Renewable Energies, headed by Dr. Wolfgang Palz. Palz was a pioneer in European exploration of the possible uses of renewable energy and provided an important link to the work on the continent on solar energy, solar architecture, wind energy, biomass, and ocean energy. He went on to serve as Chairman of the World Council on Renewable Energy (WCRE).

Wolfgang Palz

We also established contact with Hermann Scheer, a member of the German Parliament. He led the fight in Germany for a feed-in tariff (FIT) to incentivize the deployment of renewable energy technologies. The FIT system subsequently led to Germany becoming the world's leading nation in solar energy deployment. He also served as President of the European Association for Renewable Energy (EUROSOLAR).

In addition to my interest in advancing the global development of renewable energy, I was also motivated by the prospect of large potential markets for U.S.-manufactured technologies. The China program was built on a formal 1996 agreement between the two countries that called for cooperation on solar, wind, and biomass as well as training in business procedures. It was to evaluate progress on these agreements that I took my first trip to China in 1998. In a 10-day visit involving six cities, we were exposed to relevant Chinese activities and met officials responsible for Chinese energy programs. It gave me an appreciation for Chinese concern about environmental pollution, their commitment to clean energy technologies, their honesty in evaluating their own progress, and their clear interest in learning from others. In fact, on my first day in Beijing, at a meeting with the Chinese Ministry of Science and Technology, I was struck by the fact that even as I was just sitting down to start the meeting, my Chinese host began asking about electric vehicles. Beijing had recently started using cars in addition to bicycles for getting around the city, much greater use of cars was clearly anticipated, and pollution from car exhausts was already adding to their existing pollution burden from coal-burning power plants. They were clearly looking for solutions.

Shortly after my return from this trip, I was invited to underwrite the Chinese geothermal heat pump demonstration project that I described in Chapter 7 and write the preface for a Chinese progress report on their photovoltaics program. The report was published in 1999 in both Chinese and English and in that preface I stated: "With a degree of honesty not often seen, it critically examines China's past and current efforts to use photovoltaics and develop a PV industry and draws useful lessons for the future.... China's resources, population and need potentially could make it the world's largest developer of PV in the future, if appropriate policies are implemented and the investment environment is attractive." The Chinese obviously followed through.

Another interesting follow-up to my trip was a return visit by a Chinese delegation to the United States about a month later. At this get-together, I was invited to make opening remarks and I stated what I now considered the obvious: "The Chinese are more capitalistic than Americans." I stand by that statement today.

The program with India started about a year after Secretary of Energy Hazel O'Leary's famous (some would say infamous) trip in which she brought a large delegation of private sector businessmen and a few staff to New Delhi in a chartered aircraft to explore market possibilities. That trip had major ramifications. As mentioned in Chapter 9, it apparently struck the new Republican leadership in Congress as excessive and resulted in Congress making clear its resistance to DOE clean energy investments in cooperation with other countries. This resulted in the gun-shy EERE leadership discouraging such expenditures, although not putting an absolute ban on them.

A year after Secretary O'Leary's trip, I traveled to New Delhi as part of a DOE delegation to discuss a range of energy issues, including fossil fuels and renewables. It was striking how well the Secretary's trip was remembered when we arrived in Delhi— she had a colorful personality that impressed the Indians greatly. With regard to renewables, India had impressed the world by establishing a renewable energy ministry, a global first, and I was invited on the trip to share information on our programs and learn about Indian programs and possibilities for cooperation. For budgetary reasons we were unable to do as much with India then as we did with China, but India has continued its strong interest in renewables and promises to be a leading manufacturer, user, and exporter of wind and solar technologies in the future.

One interesting side note about our interaction with India. A while after my trip, the Indian renewable energy Minister undertook a return trip to the United States. He was scheduled to start with a stop in Golden, Colorado, at the NREL before a visit to my office in DC. The week before the visit, Pakistan had set off a nuclear bomb test explosion, and in response India set off their own on the weekend before the return visit. On Monday morning, while the Indian delegation was still at the NREL, I received a message from our National Security Agency (NSA) that I could not receive the Indian Minister because of U.S. policy that imposed sanctions on India for its nuclear test. In one of the stranger phone calls of my career, I had to call the Indian Minister at the NREL, inform him of the U.S. decision, and inform him that the NREL security personnel would be accompanying his delegation off the site. He understood, did not make a fuss, and returned to India.

Writing about this incident reminds me of another strange interaction with foreign visitors. In 1997 I received word that a group from North Korea wanted to visit my office to discuss renewable energy technologies. I set up the meeting, which was well within U.S. policy guidelines despite the difficulties in the U.S.–North Korean relationship. Then, as the meeting loomed and the North Koreans were in the United States, a serious conflict in my schedule arose and I proposed cancelling the meeting. Shortly thereafter I heard from the NSA that I had to hold that meeting—the NSA badly wanted pictures of the North Korean delegation and such pictures were a regular feature of such visits. The rationale I was given was that North Koreans sometimes traveled under false names and the NSA wanted to make sure they knew who was visiting the United States. Of course, being a patriotic American, I kept the meeting and got the desired photographs.

Author (center) with several U.S. colleagues and the North Korean delegation

One other unusual foreign interaction comes to mind. We regularly had Japanese visitors at the DOE and a Japanese renewable energy delegation requested a meeting with me on December 7. I would normally have scheduled the meeting as my calendar was clear, but I still carry feelings about December 7 as the day in 1941 that the Japanese attacked Pearl Harbor, and I put off the meeting until the next day. I just couldn't do it on the 7th.

The interactions with Israel involved both solar and wind energy. Israeli engineers and entrepreneurs were early pioneers in concentrating solar power and it was an Israeli-American firm, Luz, that installed the 354 MWe of parabolic trough CSP technology in Southern California in the mid to late 1980s. Israel had fledgling programs in PV when I first met with them to discuss cooperation and they were looking for financial support for their efforts. Budget considerations precluded such support but I did establish working relationships with the relevant Israeli government agencies, took two trips to Israel to discuss our solar programs, and begin discussions on the application of solar to desalination, an important Israeli interest (see Chapter 8).

The interaction on wind energy was a bit surprising since Israel does not have significant wind resources. Their interest arose from a proposal by one of their scientists, Professor Dan Zaslavsky, to build a one kilometer high tower, spray seawater at the top which would evaporate and cool air that would fall down the inside of the tower, creating an artificial wind. Turbines placed at exhaust openings at the bottom of the tower would convert the kinetic energy of this moving air into electricity.

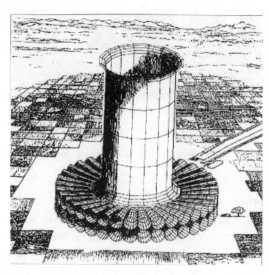

Proposed Israeli Wind Tower

The issue was complicated by the fact that an American investor in California with good political contacts supported the concept and was pushing for DOE support. I asked the NREL's wind program to evaluate the proposal and they concluded that, aside from the difficulties associated with building a one kilometer high structure and dealing with the salt that would be separated from the seawater and dispersed in the area around the tower, the resulting energy costs would not be competitive. Needless to say, I came under some political pressure for this conclusion. Continued pressure from within Israel on its own government led to a subsequent Israeli government review of the proposal as well, with the same result. In the intervening years, I have heard no further discussion of this idea.

Interactions with other countries, on a smaller scale, included meetings with Saudi Arabia (solar), Russia (solar), Jordan (solar), Japan (solar), Mexico (solar), Venezuela (geothermal), Bangladesh (solar), South Africa (solar), and Burkina Faso (solar).

An important avenue for international cooperation beyond these country-to-country interactions, and the very useful discussions with our European counterparts in the EU and Germany, was our membership and active participation in the Renewable Energy Working Group of the International Energy Agency (IEA). I represented the United States at the semiannual meetings of the 13-member Working Group from 1991–1997, most of whose members were from Europe. The multi-national IEA had been created after the OPEC-imposed Oil Embargo of 1973–1974 to ensure petroleum sharing in the event of another embargo and its principal focus was on petroleum issues.

The Working Group's focus on renewable energy was a bit of a departure for the IEA and in the 1990s it was not easy to get the IEA's attention on such matters. Nevertheless, the Working Group had a talented and committed membership and cooperative programs with limited budgets were underway on solar, wind, biomass, hydropower, geothermal, and ocean energy. The Working Group was an excellent vehicle for monitoring global progress in renewables and for information exchange. Two of its more creative activities were pioneering studies on the environmental impacts of renewable energy use and the requirements and consequences of a major deployment of PV technologies. One of my final actions as the U.S. representative was a presentation to the IEA's senior management on the potential benefits of increased use of renewable energy. It was clear that I had a somewhat skeptical audience at that time, but today the IEA is a strong supporter of clean energy efforts.

Finally, let me mention my interactions with the World Bank, which is dedicated to improving living conditions in developing countries through targeted loans. In 1993, at a meeting in Stockholm of the Renewable Energy Working Group, there was a presentation by Dennis Anderson, a senior economist at the Bank. He talked about his intent to generate enhanced interest in the Bank for supporting renewable energy projects. Up until that time, the Bank's primary energy interests were large-scale hydropower and coal-powered generation. Some efforts had started to address sustainable/renewable energy opportunities in developing regions of the world, with support from the Netherlands and the United States—e.g., the establishment in the Bank of ASTAE, the Asia Alternative Energy Unit under Loretta Schaeffer. Impressed by Dennis' talk, I contacted him upon my return to DC and made him an offer that was hard for him to refuse. Knowing that his office lacked adequate staff resources to push his renewable energy agenda within the Bank, I offered to detail a senior NREL staffer to the Bank for two years at my office's expense. I also knew who I wanted in that job, Ron White, who had worked closely with me in the DOE during my Carter Administration years, and who was now an analyst at the NREL.

Dennis was agreeable, Ron was willing, and soon Ron moved to the Bank's headquarters in DC to work with Dennis and his small staff. This was an important step forward as the Bank was a major funder of developing country activities and I wanted them favorably disposed toward clean energy technologies. Unfortunately, this movement of the Bank took longer than we anticipated, and it was frustrating to see the Bank resist this new path through much of the 1990s and early part of the new century. Today it is a different story and the Bank is now openly supportive of renewables, as is true of other international development banks.

Chapter 11

The George W. Bush–Dick Cheney Years

In 2000 George W. Bush succeeded Bill Clinton as President and the Bush–Cheney Administration came in with their own views on energy policy. Their focus was on fossil fuels and nuclear power, as it had been in the Reagan Administration. It was a discouraging period for renewables and I believe the United States lost valuable time while the rest of the world began to make significant progress in their development and deployment of renewables. We clearly lost out on the economic activity and jobs that were going to other countries as the new, clean energy industries were being established.

Vice President Cheney and President Bush

My situation also changed with the appointment of a Republican political appointee as EERE Assistant Secretary. It

The U.S. Government & Renewable Energy: A Winding Road
Allan R. Hoffman
Copyright © 2016 Allan R. Hoffman
ISBN 978-981-4745-84-0 (Paperback), 978-981-4745-85-7 (eBook)
www.panstanford.com

was mutually agreed that I would distance myself from the DOE scene of action for a while and accept a two-year detail, 2001–2003, as Senior Advisor to Winrock International's Clean Energy Group, with responsibility for water-energy issues. Winrock International is "a nonprofit organization that works with people in the U.S. and around the world to empower the disadvantaged, increase economic opportunity, and sustain natural resources." In 2014 it was responsible for 209 projects in 50 developing countries. In my two years at Winrock, I was able to bring a strong understanding of renewable energy options to their water activities, which are supported largely by USAID. I also used the time at Winrock to deepen my understanding of the water-energy nexus and try to spread appreciation of its importance.

A particularly important project was on an island in the Philippines isolated from the mainland and the electricity grid. Fresh water was brought in by boat. While fresh water existed in an underground aquifer on the island there was no way to extract it without electrically powered pumps. The solution was to use solar panels to provide the electricity to pump the water as well as to provide electricity for other purposes. The issue was how to pay for the panels—the islanders were poor and a poor risk to pay for their services. The ingenious solution was to provide water and electricity via metered services—you paid up front in incremental amounts for what you wanted. It was also recognized that people valued water more than electricity so the rates for water were set a bit high, knowing that people would pay, effectively subsidizing the electricity service. It worked very well.

Upon my return to the DOE in 2003, I was assigned to EERE's Office of Policy and Budget where I served as sort of an elder statesman with considerable responsibility for my own calendar. I advised on R&D needs and associated budget requirements as needed, mentored some young interns, and continued learning, writing, and speaking about renewable energy and water-energy issues. I was also part of a group of senior-level program officials from DOE headquarters and the DOE national laboratories that met regularly at one-day get-togethers in DC to discuss the DOE's R&D needs. This group had been meeting since the 1990s but had never focused on water-energy issues, so I encouraged the meetings' organizer, a retired official from Oak Ridge National

Laboratory, to give me afternoon time at one of the meetings to discuss the water-energy connection. He agreed, although personally skeptical of national laboratory interest in the subject. He also agreed to my request to allow 5 minutes for each interested laboratory representative to comment on the issue. I did an opening overview and when the floor was turned over to the lab representatives we eventually ran out of time because of the strong interest expressed. It quickly became clear that many laboratories had been quietly looking at water-energy issues and that greater coordination of those efforts was needed. As mentioned briefly in Chapter 8, at the end of the meeting lab representatives met at the back and organized a national laboratory partnership that continues to this day.

I also agreed to assist the DoD in reviewing energy proposals that were beginning to overwhelm their technical review capabilities. Specifically, I was working with two DoD programs, the Strategic Environmental Research and Development Program (SERDP) and the Environmental Security Technology Certification Program (ESTCP) that are jointly managed on behalf of the Office of the Deputy Under Secretary of Defense (Installations and Environment).

DoD's Environmental Research Programs

The joint office has responsibility for the entire range of scientific and technical issues associated with environmental problems associated with DoD activities, from basic research through demonstration. They routinely accept proposals annually from military-industrial teams seeking DoD support, and the increasing focus on energy issues at military installations had given rise to many proposals that were beyond the SERDP/ESTCP staff's review capabilities. The DoD contacted the DOE for help and, after some discussion, I ended up working with the joint office for three years. I was the only DOE representative, along with representatives from the various military services and the Office of the Secretary of Defense. It exposed me to the strong and growing interest in the DoD to energy and water issues and allowed me to contribute a needed expertise. Primary focus was on solar, wind, and geothermal heat pump proposals, energy and water conservation proposals, and mini-grid applications for military bases. It reinforced my earlier belief in the DoD being an important partner to the DOE in moving toward a clean energy society.

Given the flexibility of my position in the EERE Office of Policy and Budget I was able to pursue my interests in the water-energy nexus and was a frequent author and invited speaker on the subject. With the election of Barack Obama as President in 2008 there was increased attention to climate change issues in the Executive Branch. This included issuance of an Executive Order that required the federal departments and agencies to study the issue in depth, identify potential impacts on federal activities, and prepare recommendations. This led to the formation of many interagency committees, and since changes in water availability is an important impact of climate change, and energy impacts are central to global warming, I represented the DOE on several of these committees. These efforts showed that different parts of the federal bureaucracy can work together productively and a detailed and useful report was delivered to the President within a few months.

As mentioned earlier, support for renewable energy was not a priority of the George W. Bush Administration. Thus it was surprising to many of us when President Bush emphasized his support for the DOE's hydrogen program in an early State of the Union address, and recommended increased budget support. However, when his budget request was submitted to Congress a short time later, it became clear where the additional funds would come from, the other renewable energy budgets. Those of us enthusiastic about hydrogen and its use in fuel cells as a long-term energy resource, and its use as an energy storage medium, were nevertheless disheartened by this development. In my opinion it reflected the Administration's lack of interest in moving forward on renewable energy, and even a desire to slow its development down. A perhaps apocryphal story was told to me by someone who attended a meeting at the White House between the President and a senior level energy delegation from the Philippines. The head of the delegation told the President about

Philippine efforts on solar energy and the President's response was to ask why they were doing so when solar energy doesn't really work.

While the 2008 election put Barack Obama into the White House, and I was an Obama supporter, it created a problem for me. I was then 71 years old and giving serious consideration to retiring. Friends had been asking me for years why I hadn't already retired and I was getting tired of my usual answer: I'm still having fun doing what I do. Energy had been a principal focus of most of my professional life and much still remained to be done in moving the United States to a clean energy path. Why retire when the United States finally had a President who really seemed to "get it" and was supportive of what I had been advocating for many years. In fact, I sat down one Tuesday afternoon in October in a reflective mood, just before the election, and started to write down what I would say if asked what should the energy policy of the U.S. be? My answer, which I finished writing the next morning, was published in late November in the *e-journal RenewableEnergyWorld.com* and is reproduced below. My decision about retirement is discussed in Chapter 12.

Thoughts on an Energy Policy for the New Administration

While there are many divergent views on an appropriate U.S. energy policy, it is perhaps helpful to start by identifying "facts" on which most can agree:

- People do not value energy, they value the services it makes possible—heating, cooling, transportation, etc. It is in society's interest to provide these services with the least energy possible, to minimize adverse economic, environmental and national security impacts.
- Energy has always been critical to human activities, but what differentiates modern societies is the energy required to provide increasingly high levels of services.
- Population and per capita consumption increases will drive increasing global energy demand in the 21st century. While not preordained, this increase will be large even if others do not achieve U.S. per capita levels of consumption.
- Electrification increased dramatically in the 20th century and will increase in the 21st century as well. The substitution of electricity for liquid transportation fuels will be a major driver of this continued electrification.

- Transportation is the fastest growing global energy consumer, and today more than 90% of transportation is powered by petroleum-derived fuels.
- Globally energy is not in short supply—e.g., the sun pours 6 million quads of radiation annually into our atmosphere (global energy use: 460 quads). There is considerable energy under our feet, in the form of hot water and rock heated by radioactive decay in the earth's core. What is in short supply is inexpensive energy that people are willing to pay for.
- Today's world is powered largely by fossil fuels and this will continue well into the 21st century, given large reserves and devoted infrastructure.
- Fossil fuel resources are finite and their use will eventually have to be restricted. Cost increases and volatility, already occurring, are likely to limit their use before resource restrictions become dominant.
- Increasing geographic concentration of traditional fossil fuel supplies in other countries raises national security concerns.
- The world's energy infrastructure is highly vulnerable to natural disasters, terrorist attacks and other breakdowns.
- Energy imports, a major drain on U.S. financial resources, allow other countries to exert undue influence on U.S. foreign policy and freedom of action.
- Fossil fuel combustion releases CO_2 into the atmosphere (unless captured and sequestered) which mixes globally with a long atmospheric lifetime. Most climate scientists believe increasing CO_2 concentrations alter earth's energy balance with the sun, contributing to global warming.
- Nuclear power, a non-CO_2 emitting energy source, has significant future potential but its widespread deployment faces several critical issues: cost, plant safety, waste storage, and weapons non-proliferation.
- Renewable energy (solar, wind, biomass, geothermal, ocean) has significant potential for replacing our current fossil fuel based energy system. The transition will take time but we must quickly get on this path....

Recommendations

- Using the bully pulpit, educate the public about energy realities and implications for energy, economic and environmental security.

- Work with Congress to establish energy efficiency as the cornerstone of national energy policy.
- Work with Congress to provide an economic environment that supports investments in energy efficiency, including appropriate performance standards and incentives, and setting a long-term, steadily increasing, predictable price on carbon emissions (in coordination with other countries). This will unleash innovation and create new jobs.
- Consider setting a floor under oil prices, to insure that energy investments are not undermined by falling prices, and using resulting revenues to address equity and other needs.
- Work with Congress to find an acceptable answer to domestic radioactive waste storage, and with other nations to address nuclear power plant safety issues and establish an international regime for ensuring nonproliferation.
- Establish a national policy for net metering, to remove barriers to widespread deployment of renewable energy systems.
- Provide incentives to encourage manufacture and deployment of renewable energy systems that are sufficiently long for markets to develop adequately but are time limited with a non-disruptive phaseout.
- Aggressively support establishment of a smart national electrical grid, to facilitate use of renewable electricity anywhere in the country and mitigate, with energy storage, the effects of intermittency.
- Support an aggressive effort on carbon capture and sequestration, to ascertain its feasibility to allow continued use of our extensive coal resources.
- Remove incentives for fossil fuels that are historical tax code legacies that slow the transition to a new, renewables-based, energy system....

In July 2014 I revisited the October 2008 article to see what had changed in the intervening six years and to see if I would change any of the "facts" and recommendations I had put forth. One immediate reaction, which I expressed in a blog post, was "What I find interesting about this piece is that I could have written it today and not changed too many words, an indication that our country is still struggling to define an energy policy."

In my opinion three things had changed in my tabulation of "facts": the availability of large amounts of home-grown natural gas and oil at competitive prices via hydraulic fracturing (fracking) of shale deposits has turned the U.S. energy picture upside down. It may have that effect in other countries as well. Whereas the U.S. was importing over 50% of its oil just a few years ago, that fraction is now under 40% and the U.S. is within sight of becoming the largest oil producer in the world, ahead of Russia and Saudi Arabia. Whereas in recent years the U.S. was building port facilities for the import of LNG (liquified natural gas) these sites are being converted into LNG export facilities due to the glut of shale gas released via fracking and the large potential markets for U.S. natural gas in Europe and Asia where prices are higher than in the U.S.

The phenomena of global warming and climate change due to mankind's combustion of carbon-rich fossil fuels are also becoming better understood, climate change deniers have become less visible, and the specific impacts of climate change on weather and water are being actively researched. An important change is the substitution of natural gas for coal in new and existing power plants, which has reduced the share of coal from 50% just a few years ago to less than 40% today. This has reduced U.S. demand for domestic coal, which is now increasingly being sold overseas.

I still support the list of recommendations, buttressed by the following observations:

- more public education on global warming and climate change has taken place in recent years, and a majority of Americans now accept that global warming is driven by human activities.
- there is a lot of lip service given to the need for increased energy efficiency, and President Obama's agreement with the auto industry to increase Corporate Average Fuel Economy (CAFE) standards over the next decade is an important step forward. What is lacking, and slowing needed progress toward greater efficiency, is a clear policy statement from the U.S. Congress that identifies and supports energy efficiency as a national priority.
- with the shutting down of the Yucca Mountain long-term radioactive waste storage facility in Nevada, the Obama Administration is searching for alternatives but believes the country has time to come up with a better answer. This may be true, or may not, and only time will tell. It is not a uniquely American problem—other countries are

struggling with this issue as well and most seem to favor deep geological storage. This is a problem we will definitely be handing down to our children and grandchildren,

- net metering as a national policy, as is true in several other developed countries, has gone nowhere in the six years since 2008. It is another example of a lack of Congressional leadership in establishing a forward-looking national energy policy.

- progress has been made on moving renewable energy into the energy mainstream, but we have a long way to go. NREL's June 2012 report entitled 'Renewable Electricity Futures Study' made it clear that renewables could supply 80% of U.S. electricity by 2050 if we have the political will and make appropriate investments. The study puts to rest the argument used by the coal and other traditional energy industries that renewables can't do the job. The public needs to understand that this argument is inaccurate and not in our country's long term interests.

- the need for a national grid, and localized mini-grids (e.g., on military bases), has been recognized and appropriate investments are being made to improve this situation. A national smart grid, together with energy storage, are needed to assure maximum utilization of variable clean energy sources such as wind and solar. Other renewable energy sources (geothermal, biomass, hydropower, ocean energy) can be operated as baseload or near base load capacity. And even intermittent wind and solar can supply large amounts of our electricity demand as long as we can transfer power via the national grid and use averaging of these resources over large geographical areas (if the wind isn't blowing in location X it probably is blowing in location Y).

- the carbon capture and sequestration effort does not seem to be making much progress, at least as reported in the press.

- with respect to reducing long-standing and continuing subsidies for fossil fuel production, no progress has been made. Despite President Obama's call for reducing or eliminating these subsidies Congress has failed to act and is not likely to in the near-term future. This is a serious mistake as these industries are highly profitable and don't need the subsidies which divert public funds from incentivizing clean energy technologies that are critical to the country's and the world's energy future.

- today's electric utility sector is facing an existential threat that was not highly visible just a few years ago. This threat is to the utility sector's 100 year old business model that is based on generation from large, centralized power plants distributing their energy via a radial transmission and distribution network. With the emergence of low-cost decentralized generating technologies such as photovoltaics (PV), these business models will have to change, which has happened in Germany and will eventually happen in the U.S.

To close this chapter, I would just repeat my strong belief that we need to put a long-term, steadily increasing price on carbon emissions that will unleash private sector innovation and generate revenues for investments in America's future. This is a critical need if we are to successfully address climate change, create new U.S. jobs in the emerging clean energy industries, and set an example for the world.

Chapter 12

Time in the Obama Administration and Final Years in Government

In 2008, with the election of Barack Obama to succeed George W. Bush as President, and beginning my eighth decade of life, I had a decision to make. It would have been a natural time to move on to the next and presumably easier phase of my life at the end of one Administration and the onset of another, and it looked like my wife and I would be OK financially if I retired from the DOE. The following considerations guided my decision: It was getting to be a bit less fun to be at the DOE, especially during the Bush–Cheney years, but I had managed to use my seniority-enabled flexibility to pursue new intellectual challenges. I still looked forward to going to work every day and sharing my perspectives, knowledge, and experience with younger colleagues. In fact, it was at this time I formulated what I called the Hoffman Rule, that the first two thirds of one's life should be devoted to learning and experiencing and the last third to sharing that knowledge with people who are younger and will come after you. This is the rule I followed when I finally did retire in 2012, at the age of 75 and continue to follow. Hence, my blog Web site, which now holds more than 100 blog posts, this book, and a number of volunteer advisory activities.

With Obama's election I gave careful consideration to staying at the DOE, using the rationale that I had worked for the past 39

The U.S. Government & Renewable Energy: A Winding Road
Allan R. Hoffman
Copyright © 2016 Allan R. Hoffman
ISBN 978-981-4745-84-0 (Paperback), 978-981-4745-85-7 (eBook)
www.panstanford.com

years to advance the concept and practicability of clean energy technologies, and why would I retire just when we finally had a President who really seemed to "get it" in a meaningful way. I was also enjoying my work with the DoD and represented the DOE on several interagency committees beginning to focus on global warming and climate change, an issue I was deeply interested in. And of course, I could retire at any time if I so decided. The choice was mine.

President Obama and Vice President Biden

What I hadn't counted on was the behavior of the newly appointed leadership of EERE. My role in the Office of Policy and Budget wasn't a great fit, given my broad interests and experience, but it came with a welcome degree of flexibility. The new leadership, people I knew well, had a different plan. I was to be transferred to the wind program, which could use some graybeard help, but they failed to discuss this with me for several months after they assumed their responsibilities. I found out indirectly during a routine performance review with my direct supervisor, who was surprised that I didn't know. Shortly thereafter I walked by a friend in the hallway who headed up the wind program, and in passing he said: "I hear you're going to join us." At this point I simply picked up the phone and called the wind program to find out what was going on. They were surprised as well that I had been kept in the dark about the putative move. After this discussion, which confirmed the EERE front office's intentions, I sat down and wrote a long memo to our new Assistant Secretary and his Deputy, expressing my dismay at their failure to communicate, laying out the full range of my responsibilities for which there was no backup in the organization, and requesting a meeting. That meeting with the Deputy took place about a week later and it quickly became clear that he had not read my memo. I suggested we terminate the meeting until he had a chance to read the memo and reschedule for another time. That other time never occurred.

Faced with this situation, I set up several meetings with the wind program leadership, to see what they had I mind if I agreed to join them. One suggestion was to chat with the recently hired young head of the offshore wind (OSW) program which had just been established and needed additional personnel. His experience was in naval architecture and he had an impressive military background. In my subsequent discussion with him, I learned what he had in mind for someone with my background and experience, I liked his directness, and given my enthusiasm for offshore wind as the most important emerging renewable electric technology, I decided not to retire but to contribute to and help shape this new program. It was a good decision, but one condition was that I had to give up my work with the DoD and with water-energy nexus issues.

The next two plus years were devoted mostly to learning as much as I could about offshore wind, helping to plan the new program, promote offshore wind to utility and other outside audiences through my writing and speaking, and mentor the younger members of the OSW staff. These were fun things for me to do, and thoughts of retirement receded into the background. What finally changed my mind were increasing frustration with the wind program's leadership, the departure of the OSW program head under difficult circumstances, my growing realization that the fun in my job was disappearing, and that I was getting older and had things I wanted to do in retirement, mainly involving writing. Thus, after many discussions with my wife and a lengthy internal process I decided to retire at the end of December 2012. However, before I could announce my retirement, and at a time of tight budgets, the DOE announced a buy-out program to reduce its workforce and I took the buy-out, which advanced my retirement by three months as I was required to leave the payroll by September 30, the end of the 2012 fiscal year.

Let me say a few words about staff development and mentoring, which I have always considered important parts of my job. It may be defined as a relationship in which a more experienced or more knowledgeable person helps to guide a less experienced or less knowledgeable person. The mentor may be older or younger, but have a certain area of expertise.

Staff development is critical to organizational success and requires consistent top down encouragement and support. Lots of organizations talk about staff development as a worthy goal, and there is an extensive literature on the subject, but truly successful programs are limited in number because of lack of follow-through and true organizational commitment to the goal. Without senior management buy-in, and publicly expressed support for such an effort, success is unlikely to happen, given everyone's assigned responsibilities and management's focus on managing the latest crisis, often referred to as "firefighting."

Now what about mentoring. People are an organization's most important asset, and it is through people that we can have lasting impact on that organization. You make an enduring difference through the people you choose to develop. In addition to supporting continuing education, a critical ingredient of staff development is providing a mentor, someone whose knowledge and experience the mentee respects and someone whose wisdom and knowhow can support the professional growth and development of the mentee. Corporate mentoring programs have long been recognized as an essential strategy for attracting, developing, and retaining top employees. They send a message to employees that they are valued and the organization wants them to be satisfied and happy.

Mentoring helps new employees settle into an organization, understand what it means to be a professional in their working environment, facilitates the transfer of expertise to those who need to acquire specific skills, encourages the development of leadership abilities, and helps employees plan, develop, and manage their careers. Mentoring is also a two-way street that can benefit both the mentee and the mentor.

Unfortunately, many organizations do a poor job of "preparing the next generation." This was my consistent and depressing view during my many years at the DOE. In my opinion not enough attention was paid to planning, supporting, and rewarding mentoring activities. A few other federal departments and agencies did a better job at this during my time in government—e.g., as evidenced by their heavy participation in the one-year-long, government-wide Excellence In Government Program. Such programs do take time away from other activities but are investments in the future, just as are federal R&D investments in emerging energy technologies.

A few words are also in order on my enthusiasm for offshore wind, which is today a rapidly advancing global energy technology and is becoming the principal focus of EERE's wind energy R&D program. While onshore wind is growing rapidly and there are significant opportunities for its further development, the offshore potential may be even greater. To put this statement in context, while total U.S. installed generating capacity today is just over 1,000 GWe, the gross U.S. offshore wind resource out to 50 nautical miles, prior to adjustments for shipping lanes, underwater cables, and environmental and economic constraints, has been estimated by the NREL to be more than 4,000 GWe. Thus, if even a fraction of this potential was to be realized, the United States has a massive new energy resource waiting to be tapped. Developing this resource has significant potential for revitalizing the U.S. manufacturing sector and job creation, with the NREL estimating that "...offshore wind will create more than 20 direct jobs (in addition to many indirect jobs) for every megawatt produced in the United States." Offshore wind is also an indigenous, non-depletable, zero-carbon-generating technology for which the energy resource is located close to load centers on both U.S. coasts, the Great Lakes, and the Gulf of Mexico. And finally, winds in the offshore regions have higher average speeds than onshore winds, blow more regularly, and OSW turbines can be built without size limitations. These final three considerations improve the economics of wind power since the average power extracted from the wind by a turbine scales with the third power of the wind speed, more regular winds increase wind's capacity factor, and energy costs scale inversely with turbine size. An interesting fact is that wind turbines onshore are limited to about 3 MWe because of constraints imposed by transportation limitations with large loads.

Offshore wind farms

The potential for offshore wind has been recognized broadly, and today the United Kingdom leads the world in OSW deployment. Other countries in Europe are beginning to tap this potential as well. Only China, with its long coastlines, has more OSW potential than the United States and is moving rapidly to become a world leader in this technology, as it has in the fields of onshore wind and PV. More recently Japan, in the wake of the Fukushima nuclear accidents, has also committed to an aggressive program on OSW. Unfortunately, the United States has gotten a late start on developing this technology, and it will have its first OSW steel in the ocean in 2016. All of these national programs are working hard to bring down the costs of OSW, today 2–3 times the KWh cost of onshore wind, and to develop floating OSW turbines to take advantage of the wind resources in deep waters.

Upon my departure from government service, only one task remained. A few months before my retirement, I had been invited to author a 5,000–6,000-word chapter on water-energy issues that would be included in a new book on energy poverty to be published by Oxford University Press. I had agreed to do this, it was to be the only chapter on such issues in a 22-chapter book, and I saw it as an opportunity to continue spreading the word. As soon as I "retired" (but only from my DOE job), my principal occupation, besides sleeping in a bit later, enjoying a more leisurely breakfast, assuming full responsibility for walking our dog, was to draft this book chapter. It took most of October, and subsequently a part of January 2013 when I was asked by the editors to add some additional details. The book was finally published with the title *Energy Poverty: Global Challenges and Local Solutions* in December 2014.

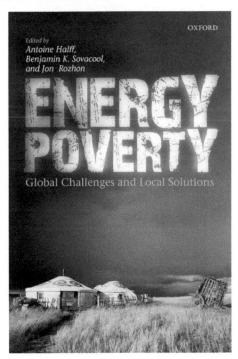

Chapter 13

Summarizing and Looking at Today's Renewable Energy Situation

In this book I have tried to describe some steps on the road from dependence on traditional energy sources, fossil fuels and nuclear, to a clean energy world increasingly dependent on renewable energy. I have been privileged to observe and participate in this transition in a relatively unique way. It has been exciting and fulfilling for me, and I am grateful that I have lived long enough to see the beginnings of this transition on a global basis. In fact, I am gratified and even amazed that every day seems to bring new evidence of this energy revolution. I feel confident that the energy world that our children will inhabit later in this century, and eventually their children and grandchildren, will be very different from the fossil-fuel dominated energy world we have today. What is painful to me about today's energy situation is that the United States is not yet leading this transition. Too much attention is still being paid in the U.S. Congress to protecting vested energy interests. This has to change if the United States is to derive the full economic, environmental and political benefits of the clearly unfolding move to a clean energy world, which admittedly will take time, as history teaches, but which I believe is inevitable.

Let me try to summarize briefly what the preceding chapters have attempted to document and review today's energy situation

The U.S. Government & Renewable Energy: A Winding Road
Allan R. Hoffman
Copyright © 2016 Allan R. Hoffman
ISBN 978-981-4745-84-0 (Paperback), 978-981-4745-85-7 (eBook)
www.panstanford.com

as I see it. In the next and final chapter, I will look ahead to what I see coming in the years and decades ahead. This book starts with the beginning of the nuclear power debate in the United States, the vehicle that got me interested in energy issues. It proceeds through the 1970s, when the 1973–1974 OPEC oil embargo brought energy to the world's attention, nuclear power was designated by some as the energy source of the future, and describes the Carter Administration's attempts to begin looking at renewable energy technologies other than hydropower via the Domestic Policy Review of Solar Energy. This was also the era when the U.S. Congress passed legislation setting fuel economy standards for new automobile fleets and when a few prescient scientists began to warn about global warming and its potential for global climate change. In energy terms the 1970s were a turning point in alerting the world to the shortcomings of a fossil-fuel dominated global energy system and the need to examine long-term alternatives. It is also the period in which an increasing number of people began to examine carefully the potential liabilities of a nuclear economy after the Chernobyl and Three Mile Island nuclear power plant accidents.

The 1980s saw a change in U.S. Administrations and a reduced interest in clean energy alternatives to traditional energy technologies. This point of view was strengthened by falling oil prices (at one point in the mid 1980s the price of a barrel of oil fell below $10), average energy costs in the United States were well below those of most other nations, and public attention to energy issues dropped precipitously. It was also the period when the Reagan Administration tried to kill the budget for renewable energy R&D and even tried to eliminate the DOE entirely. It was a bad time for renewables, salvaged only by the efforts of a few dedicated individuals. It was also a time when the big oil companies made some investments in solar energy, but this stopped when they realized that solar was likely to be a long-term investment opportunity. The end of the 1980s saw the election of George H. W. Bush as President and the beginning of a reversal in renewable energy's fortunes. R&D budgets began to recover and President Bush indicated his support for renewable energy by converting SERI into the National Renewable Energy Laboratory.

The 1990s were a period of change for renewables with the election of Bill Clinton as President and Al Gore as Vice President.

Both had familiarity with energy issues from their previous positions, understood the importance of moving to a clean energy society, and renewable electric R&D budgets reached and exceeded $300M. Global warming also began to attract considerable public attention. Unfortunately, in 1995 control of Congress passed from the Democrats to the Republicans and the Gingrich Revolution produced significant cuts in these R&D budgets to the extent that the NREL had to lay off one quarter of its staff. Nevertheless, solar and wind energy costs continued to drop and progress was made in advancing other renewable electric technologies. In fact, there was enough attention beginning to be paid to renewables that companies in the coal industry, which supplied the bulk of the fuel for U.S. electricity generation, began to attack renewables as incapable of meeting national energy needs. Storage was also recognized as an important component of a renewables-based economy, due to the variability of wind and solar, and hydrogen began to be taken seriously as an energy fuel and energy storage medium. It was also the period during which a number of other countries began to explore their renewable energy options and to develop indigenous renewable energy industries.

The turn of the century saw a new Republican Administration in the United States with a focus on fossil fuels and nuclear power and relatively little interest in renewables. However, other countries were moving ahead with their renewable energy programs, clearly recognizing the need to move away from dependence on fossil fuels for environmental, economic, and national security reasons. Incentives began to be offered to encourage this transition—for example, Germany introduced its FIT (Feed-in Tariff) program to stimulate the deployment of solar, wind, and other renewable electric technologies, and today Germany leads the world in installed solar energy systems. The UK recognized the importance of its onshore and offshore wind resources and today leads the world in offshore wind energy deployments. China undertook a major effort to develop its renewable energy resources and today leads the world in PV cell and wind turbine production. It will not be too many years before it assumes the lead in offshore wind energy deployment as well.

The new century also saw the acceptance of global warming and associated climate change as issues of global importance and requiring global cooperation. It was also the period when the inextricable linkage between water and energy issues was finally recognized, along with the other linkages among water, energy, food, environment, and public health. Both sets of issues, along with clean energy issues, are creating critical R&D and policy agendas for the 21st century.

The second decade of the 21st century also saw the beginning of the inevitable transition to a renewable energy society, both in the United States and around the world. As I write this in early 2016, I am aware of the fact that solar is the fastest growing energy source in the world today, even exceeding the rapid rate at which wind energy continues to grow and will continue to grow as additional onshore sources are exploited and offshore wind emerges. An excellent summary of renewable energy's status as of the end of 2014 has been provided by REN21 in the Executive Summary of its annual "Renewables 2015 Global Status Report," which I reproduce here in full. REN21 is an international non-profit association that facilitates "knowledge exchange, policy development and joint action towards a rapid global transition to renewable energy."

EXECUTIVE SUMMARY

Renewable energy continued to grow in 2014 against the backdrop of increasing global energy consumption, particularly in developing countries, and a dramatic decline in oil prices during the second half of the year. Despite rising energy use, for the first time in four decades, global carbon emissions associated with energy consumption remained stable in 2014 while the global economy grew; this stabilisation has been attributed to increased penetration of renewable energy and to improvements in energy efficiency. Globally, there is growing awareness that increased deployment of renewable energy (and energy efficiency) is critical for addressing climate change, creating new economic opportunities, and providing energy access to the billions of people still living without modern energy services. Although discussion is limited to date, renewables also are an important element of climate change adaptation, improving the resilience of existing energy systems and ensuring delivery of energy services under changing climatic conditions. Renewable energy provided an estimated 19.1% of global final energy consumption in 2013, and growth in capacity and generation continued to expand in 2014. Heating capacity grew at a steady pace, and the production of biofuels for transport increased for the second consecutive year, following a slowdown in 2011–2012. The most rapid growth, and the largest increase in capacity, occurred in the power sector, led by wind, solar PV, and hydropower. Growth has been driven by several factors, including renewable energy support policies and the increasing cost-competiveness of energy from renewable sources. In many countries, renewables are broadly competitive with conventional energy sources. At the same time, growth continues to be tempered by subsidies to fossil fuels and nuclear power, particularly in developing countries. Although Europe remained an important market and a centre for innovation, activity continued to shift towards other regions. China again led the world in new renewable power capacity installations in 2014, and Brazil, India, and South Africa accounted for a large share of the capacity added in their respective regions. An increasing number of developing countries across Asia, Africa, and Latin America became important manufacturers and installers of renewable energy technologies. In parallel with growth in renewable energy markets, 2014 saw significant advances in the development and deployment of energy storage systems across all sectors. The year also saw the increasing electrification of transportation and heating applications, highlighting the potential for further overlap among these sectors in the future.

Power: more renewables capacity added than coal and gas combined Renewables represented approximately 58.5% of net additions to global power capacity in 2014, with significant growth in all regions. Wind, solar PV, and hydro power dominated the market. By year's end, renewables comprised an estimated 27.7% of the world's power generating capacity, enough to supply an estimated 22.8% of global electricity. Variable renewables are achieving high levels of penetration in several countries. In response, policymakers in some jurisdictions are requiring utilities to update their business models and grid infrastructure. Australia, Europe, Japan, and North America have seen significant growth in numbers of residential "prosumers"—electricity customers who produce their own power. Major corporations and institutions around the world made substantial commitments in 2014 to purchase renewable electricity or to invest in their own renewable generating capacity. Heating and Cooling: slow growth but vast potential—key for the energy transition About half of total world final energy consumption in 2014 went to providing heat for buildings and industry, with modern renewables (mostly biomass) generating approximately 8% of this share. Renewable energy also was used for cooling, a small but rapidly growing sector. The year saw further integration of renewables into district heating and cooling systems, particularly in Europe; the use of district systems to absorb heat generated by renewable electricity when supply exceeds demand; and the use of hybrid systems to serve different heat applications. Despite such innovations and renewables' vast potential in this sector, growth has been constrained by several factors, including a relative lack of policy support. Transport: driven by biofuels, with e-mobility growing rapidly in the transport sector, the primary focus of policies, markets, and industries has been on liquid biofuels. The share of renewables in transportation remains small, with liquid biofuels representing the vast majority. Advances in new markets and in applications for biofuels— such as commercial flights being fueled by aviation biofuel— continued in 2014. Relatively small but increasing quantities of gaseous biofuels, including bio-methane, also are being used to fuel vehicles. Increased electrification of trains, light rail, trams, and both two- and four-wheeled electric vehicles is creating greater opportunities for the integration of renewable energy into transport."

An even more recent report by the Frankfurt School-UNEP Collaborating Centre for Climate & Sustainable Energy Finance, "Global Trends in Renewable Energy Investment 2016," makes it clear that the renewable energy revolution is finally and firmly underway. A summary of its key investment findings is presented below.

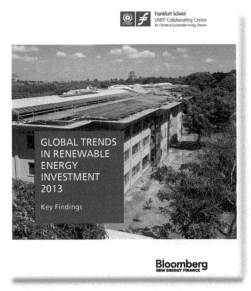

(1) 2015 produced a new record for global investment in renewable energy. The amount of money committed to renewables excluding large hydro-electric projects rose 5% to $285.9 billion, exceeding the previous record of $278.5 billion achieved in 2011.

(2) Even more striking was that the amount of generating equipment added in wind and solar photovoltaics last year came to 118GW, far above the next highest annual figure, 2014's 94GW. Overall, renewables excluding large hydro made up 53.6% of the gigawatt capacity of all technologies installed in 2015, the first time it has represented a majority.

(3) Global investment in renewable power capacity, at $265.8 billion, was more than double dollar allocations to coal and gas generation, which was an estimated $130 billion in 2015.

(4) Last year was also notable as the first in which investment in renewables excluding large hydro in developing countries outweighed that in developed economies. The developing world including China, India and Brazil committed a total

of $156 billion, up 19% on 2014, while developed countries invested $130 billion, down 8%. A large element in this turnaround was China, which lifted its investment by 17% to $102.9 billion, or 36% of the world total.

(5) However, other developing countries also raised their game—India saw its commitments rise 22% to $10.2 billion, while Brazil ($7.1 billion, down 10%), South Africa ($4.5 billion, up 329%), Mexico ($4 billion, up 105%) and Chile ($3.4 billion, up 151%) all joined it in the list of the top 10 investing countries in 2015. The list of developing countries investing more than $500 million last year included Morocco, Uruguay, the Philippines, Pakistan and Honduras.

(6) Investment in Europe slipped 21% to $48.8 billion, despite the continent's record year for financings of offshore wind, at $17 billion, up 11%. The US enjoyed a 19% bounce in renewable energy commitments to $44.1 billion, its highest level since 2011, with solar accounting for just over two thirds of that total. Japan attracted $36.2 billion, almost the same as in 2014, thanks to its continuing boom in small scale PV.

(7) Renewable generation costs continue to fall, particularly in solar photovoltaics. In the second half of 2015, the global average levelised cost of electricity for crystalline silicon PV was $122 per MWh, down from $143 in Q2 2014. Specific projects are going ahead at tariffs well below that—the record-breaker so far being a 200 MW plant in Dubai being built by ACWA Power International, awarded a contract in January 2015 at just $58.50 per MWh.

(8) Public market investment in renewable energy totaled $12.8 billion in 2015, down 21% on the previous year's figure but close to the average of the years since 2008.

(9) There is rising interest in battery storage as an adjunct to solar and wind projects and to small-scale PV systems. In 2015, some 250MW of utility-scale electricity storage (excluding pumped hydro and lead-acid batteries) were installed worldwide, up from 160MW in 2014. Announced projects reached 1.2GW.

Chapter 14

Looking Ahead

Looking ahead must be done with humility, recognizing that for most of us it is a hard thing to do well. Arthur Clarke, the distinguished engineer (he was the first to propose stationary orbit satellites) and science fiction writer, addressed this issue in his 1962 book *Profiles of the Future: an Inquiry into the Limits of the Possible*. He points out the difficulty even the most distinguished people in a field have of accurately seeing what is coming down the road. He gave several examples and proposed the Three Laws that are still quoted today:

(1) When a distinguished but elderly scientist says that something is possible, (s)he is almost certainly right. When (s)he says it is impossible, (s)he is very probably wrong.
(2) The only way of finding the limits of the possible is by going beyond them into the impossible.
(3) Any sufficiently advanced technology is indistinguishable from magic.

These are important insights that must be kept in mind.

Energy is the area where I have devoted the bulk of my professional career and where my credibility may be highest. At least I'd like to think so. Recognizing from history that changes in our energy systems most often come slowly over decades but occasionally more rapidly, these are my current thoughts on where I anticipate we will be energy-wise in 30–40 years.

The U.S. Government & Renewable Energy: A Winding Road
Allan R. Hoffman
Copyright © 2016 Allan R. Hoffman
ISBN 978-981-4745-84-0 (Paperback), 978-981-4745-85-7 (eBook)
www.panstanford.com

Arthur Clarke

Renewable energies, i.e., solar, wind, hydropower, geothermal, biomass, and ocean energy, are not new but, except for hydropower, their entering or beginning to enter the energy mainstream is a relatively recent phenomenon. Solar in the form of photovoltaics (PV) is a truly transformative technology in the sense that it will change the way we use energy in the future. Our homes will routinely use PV to provide grid-independent electricity, and solar hot water heaters will provide thermal energy for hot water production, space heating, for higher temperature industrial applications, and for powering heat-powered cooling cycles. In the case of PV, this is due to significant and continuing cost reductions for manufacture of solar panels in recent years, PV's suitability for distributed generation, its ease and quickness of installation, and its easy scalability. As soon as PV balance-of-system costs (labor, support structures, permitting, wiring) are reduced and approach current PV cell costs of about $0.5–0.7 per peak watt, I expect this technology to be widespread on all continents and in all developed and developing countries. Germany, not a very sunny country but the country with the most PV installed to date, has even had occasional summer days when half or more its electricity was supplied by solar energy. In combination with energy storage to address its variability, I see PV powering a major revolution in the electric utility sector as utilities recognize that their current business models are becoming outdated. This is already happening in Germany, where electric utilities are now moving rapidly into the solar business. In terms of the future, I would not be surprised if solar PV is built into all new residential and commercial buildings within a few decades, backed up by battery or flywheel storage. I can even anticipate the widespread use of stored hydrogen created by electrolysis of water powered by excess electricity for use in fuel cells, another transformative technology. Most buildings will still be connected to the grid as a backup, but a significant fraction of domestic electricity (30–40%) could be solar-derived by 2050. The viability of this projection is supported by the NREL June 2012 Renewable Electricity Futures Study.

Hydropower already contributes about 10% of U.S. electricity and, I anticipate, will grow somewhat in future decades as more low-head hydropower sites are developed. Pumped storage, a form of hydropower that is used as a storage medium for electric energy, will also be expanded as all forms of storage become more important parts of our electricity system.

For many years, onshore wind was the fastest growing renewable electricity source until overtaken recently by PV. It is still growing rapidly and will be enhanced by offshore wind, which is slowly emerging. However, I expect offshore wind to grow rapidly as we approach mid-century as costs are reduced for two primary reasons: It taps into an impressively large energy resource off the coasts of many countries, and it is in close proximity to coastal cities where much of the world's population is increasingly concentrated. Floating wind turbines, suitable for deployment in deep ocean waters where wind resources are most favorable, are under active development. In my opinion, wind, both onshore and offshore, together with solar and hydro, will contribute 50–60% of U.S. electricity in 2050.

An important technology complementary to solar and wind energy, which are variable at individual sites, is energy storage. Proponents of traditional energy sources, such as fossil fuels and nuclear power, often point to this variability as a reason for not moving to a renewable energy-based energy economy. This argument neglects the important role that storage is already beginning to play, and will play increasingly in the future, in allowing us to take full use of our solar and wind energy resources. Electrical storage costs are coming down steadily today, and by mid-century storage of electrical energy by utilities, businesses, and individual residences will be a ubiquitous part of our daily lives.

Other renewable electric technologies will contribute as well, but in smaller amounts. Hot dry rock geothermal wells (now called enhanced geothermal systems) will compete with and perhaps come to dominate traditional geothermal generation, but this will take time. Wave and tidal energy will be developed and become more cost effective in specific geographical locations, with the potential to contribute more in the latter part of the century. This is especially true of wave energy, which taps into a large and nearly continuous energy source. It may not be unusual by mid-century to see adjacent offshore wind and wave energy "farms" in the ocean.

Biomass in the form of wood is an old renewable energy source, but in modern times biomass gasification and conversion to alternative liquid fuels is opening up new vistas for wide-scale use of biomass as costs come down. By mid-century I expect electrification and biomass-based fuels to replace our current heavy dependence on petroleum-based fuels for transportation. This trend is already underway and may be nearly complete in the United States by 2050. Biomass-based chemical feedstocks will also be widely used, signifying the beginning of the end of the petroleum era.

I expect that other fossil fuels, coal and natural gas, will still be used widely in the next few decades, given large global resources and dedicated infrastructure. Natural gas, as a cleaner burning fossil fuel, and with the availability of large amounts via fracking, will gradually replace coal in power plants and could represent 30–40% of U.S. power generation by mid-century with coal generation disappearing.

To this point I have not discussed nuclear power, which today provides close to 20% of U.S. electricity. While I believe that safe nuclear power plants can be built today—i.e., no meltdowns—cost, permanent waste storage, and weapons proliferation concerns are all slowing nuclear's progress in the United States. Given the availability of relatively low-cost natural gas for at least several decades (I believe fracking will be with us for a while), the anticipated rapid growth of renewable electricity, and the risks of nuclear power, I see limited enthusiasm for its growth in the decades ahead. In fact, I would not be surprised to see nuclear power supplying only about 10% of U.S. electricity by 2050, and less in the future.

One point I wish to emphasize is my belief that renewable energy will power development in many parts of the developing world in the 21st century. Africa is a good example, and to quote the Director General of IRENA (International Renewable Energy Agency): "Africa is blessed with plentiful land and natural resources. Prodigious sunshine blankets the continent for much of the year, ideal conditions for solar power. Hot rocks in areas such as the Rift Valley store geothermal energy. Vast plains and mountain ranges are great sites for wind turbines while mighty rivers like the Zambezi can be harnessed for hydropower projects. Finally, biomass is abundant—all providing multiple opportunities for renewable energy production."

I will also quote from a recent speech to the Brookings Institution by U.S. Senator Chris Coons of Delaware: "From urbanization and economic growth, to public health and energy, Africa is developing at a pace that rivals nearly every other region of the world. It is truly the continent of the 21st century." What will be true of Africa in this century will also be true of many other parts of the developing world.

Chapter 15

The Importance of Energy Policy

I conclude this book by noting that the United States and indeed the world is at a crossroads when it comes to the choice on how we want to provide energy services in the future. These services include lighting, heating, cooling, transportation, clean water, communications, entertainment, and commercial and industrial activities. In my opinion the United States needs a national energy policy enunciated by Congress that recognizes the importance of moving to a low-carbon future as quickly as possible. This would involve making energy efficiency, the wise use of any energy source, the cornerstone of national energy policy, and creating an investment environment that stimulates the rapid deployment of renewable energy technologies. My recommendation is to put a long-term and steadily increasing price on carbon emissions to motivate private sector decisions to, over time, use fewer fossil fuels and more renewable energy and let markets work. An alternative approach would be to progressively tax producers and importers of carbon-based fuels.

Support for non-carbon emitting or carbon-neutral renewables is also driven by increasing awareness that while nuclear power generation does not put carbon into the atmosphere, it does create multigenerational radioactive waste disposal problems, can be expensive, raises low probability but high-consequence safety issues, and is a step on the road to proliferation of nuclear weapons' capability. Nuclear power remains an option

The U.S. Government & Renewable Energy: A Winding Road
Allan R. Hoffman
Copyright © 2016 Allan R. Hoffman
ISBN 978-981-4745-84-0 (Paperback), 978-981-4745-85-7 (eBook)
www.panstanford.com

as long as the problems listed above can be addressed successfully. My personal view is that renewables are a much better choice, especially now that it has been shown that renewable energy, in its many forms, can provide the bulk of our electrical energy needs.

Unfortunately, establishment of a national policy focused on clean energy has not been a priority of the U.S. Congress in recent years. Such a policy will allow the United States to be a leader in addressing global warming and be a major economic player in the world's inevitable march to a clean energy future. Recent actions by the Obama Administration are important steps in the right direction, such as the agreement reached with the automobile industry to increase CAFE standards through 2025, and the Administration's Clean Power Plan that requires states to reduce emissions from power plants.

Energy policy is a complicated and controversial field, reflecting many different national, global, and vested interests. Bringing renewables fully into the energy mainstream, which is just now beginning, will take time as history teaches, and the needs of developing and developed nations (e.g., in transportation) need to be addressed during the period in which the transition takes place. The critical need is to move through this transition as quickly as possible. Without clear national energy policies that recognize the need to move away from a fossil fuel-based energy system, and to a low-carbon clean energy future as quickly as possible, this inevitable transition will be stretched out unnecessarily, with adverse environmental, job-creation, and other economic and national security impacts. It is also true that the revenue generated by putting a price on carbon can be used to reduce social inequities introduced by such a tax, lower other taxes, and enable investments consistent with long-term national needs. In the United States, it also provides a means for cooperation between Republicans and Democrats, something we have not seen for several decades. It is now more than time for U.S. leaders to take this critical step.

Index

T - #0585 - 101024 - C0 - 234/156/9 - PB - 9789814745840 - Gloss Lamination